CONTENTS

COLD WAR BOYS OVERSEAS

BY THE SAME AUTHOR:

Men Behind the Medals

The Buccaneers

Men Behind the Medals – A New Selection

Shot Down and on the Run

Shot Down and in the Drink

Royal Air Force Day by Day

The RAF's First Jet Squadron

The Battle of Britain Story

The Sowreys

Buccaneer Boys

Airmen Behind the Medals

Forever Vigilant

Royal Air Force Day by Day 100

Daily Telegraph Book of Airmen's Obituaries Book Three

Buccaneer Boys 2

Beaufighter Boys

Bristol Beaufighter – Wartime Operations

COLD WAR BOYS OVERSEAS

TRUE TALES BY THOSE WHO SERVED WITH THE ROYAL AIR FORCE ABROAD

GRAHAM PITCHFORK

FOREWORD BY AIR CHIEF MARSHAL SIR RICHARD JOHNS GCB, KCVO, CBE, FRAeS

GRUB STREET • LONDON

This book is dedicated to
All those who served overseas
Remembering those who failed to return

Published by
Grub Street
4 Rainham Close
London SW11 6SS

A CIP record for this title is available from the British Library

ISBN-13: 978-1-911714-05-7

Design by Myriam Bell Design, UK

Printed and bound by Finidr, Czechia

FOREWORD

Air Chief Marshal Sir Richard Johns GCB, KCVO, CBE, FRAeS

ACM Sir Richard Johns during his Harrier days.

Graham Pitchfork's introduction to this volume is the perfect scene setter to a treasure trove of memories of overseas tours when they were part of a routine career pattern, and for the majority were eagerly anticipated. And how heartening it is to read recollections that evince personal pride in achievement and a can-do attitude that was the equal of operational challenges associated with service in the Far East, the Middle East and Germany.

As the size and character of the service has gradually diversified with the passage of time it is tempting to wallow in a trough of nostalgia, albeit an educational experience for those who like me never served in the Far East Air Force. That said, chapters eighteen to twenty-four and particularly those concerning air transport and helicopter operations merit the readers' respect and admiration. These less well-known aspects of overseas service with the RAF sit alongside the more familiar activities in other theatres and remind us of the great range in capabilities and skills required of those who served in the air and on the ground in all corners of the world.

But if happy memories of overseas tours, both on and off duty, predominate there was a cost. Many contributors mention fatal accidents that occurred during their tours. The reasons were many and varied but as Stephen Wrigley so succinctly observes, "Triumph and potential tragedy can be close neighbours in the flying world". So how appropriate that this book is dedicated to "All those who served overseas remembering those who failed to return".

I very much hope that it gains the service and public acclaim it so richly deserves.

ACKNOWLEDGEMENTS

BOOKS OF THIS NATURE ONLY come into being because a bunch of chums are prepared to delve into their memory banks, recall their experiences and then spend time putting pen to paper. Many have long been retired so committing time and effort to respond to my requests has probably been an intrusion into their otherwise orderly and peaceful lives. Their names appear with their own chapter, and I trust they will accept this as due recognition and appreciation for their fascinating narratives. I am immensely grateful to all of them for 'answering the call'.

I am particularly grateful to Air Chief Marshal Sir Richard Johns, who put me through my paces as a junior flight cadet at Cranwell and who went on to be a chief of the air staff, for his eloquent foreword. With service in Aden and in Germany, he too is an 'Overseas Boy' and well versed in the many experiences we enjoyed and endured in RAF life overseas.

Most of the photographs have been provided by the authors and from my own collection. Those from other sources are credited with the photograph. I am indebted to Lee Barton at the RAF Historical Branch for his time and for finding photographs to fill any gaps.

Finally, I must thank the Grub Street team. Over many years they have given me a great deal of support in previous endeavours, and they continue to do so. Thanks to John Davies, Natalie Parker and Debbie Marr.

INTRODUCTION

For almost twenty-five years after the end of World War Two, the Royal Air Force maintained a worldwide presence, largely in those areas where the service had been based during the inter-war years. Predominately, these were in the Middle East and the Far East. The withdrawal from India in 1947, followed by Iraq in 1955 and Egypt the following year heralded a steady reduction of the RAF presence on permanent bases until the final exit from Singapore, Aden and the Persian Gulf in the late 1960s. A modest RAF presence on Cyprus continues to this day, but the large and active bases in the Mediterranean have long gone. An exception to those withdrawals from old colonial bases has been the RAF presence in Germany until 2001.

During the 1960s, almost half the RAF personnel based abroad were serving in Germany. Over 50,000 service personnel served on the continent, a figure that far exceeds the current strength of today's RAF. By the mid 1960s, the numerous operational bases had been reduced to four, but each had populations of 7–8,000 personnel including families. The RAF deployed squadrons of Hunters, Javelins and Canberras. By early 1970 these had been replaced by the new generation of combat aircraft, the Buccaneer, Harrier, Jaguar and Phantom. Helicopters were based at the forward airfield of Gütersloh and missile defences were boosted. Tornados were to follow.

With the collapse of the USSR and demolition of the Berlin Wall, so the Cold War gradually came to an end and the requirement for aircraft to be on alert at very short notice disappeared and the need for a large air force presence in Germany declined. Squadrons began to re-deploy to bases in the United Kingdom, some were disbanded and on 4 September 2001, the last Tornado departed from Brüggen, the only remaining RAF operational base.

The reason for the RAF presence on bases overseas was to meet national and international commitments. But there was much more to life overseas than the 'day job'. The large bases were more akin to mini-towns and family and socialising were a major part of daily life. Personnel had a great desire to 'serve overseas', which gave the opportunity to experience different cultures and ways of life, travel to otherwise inaccessible areas, and all with the attraction of some financial benefits.

Serving overseas was a different way of life and the chapters that follow illustrate the many facets of the RAF's presence across the globe. The wide variety of accounts have been carefully chosen to provide a graphic example of the wide-ranging, and variable, capabilities of the RAF to meet many different political and military scenarios. They also illustrate how much fun we all had and if our memories have slightly dulled since those unique and halcyon days, then so be it. Digging into the memory bank of those days fifty-odd years ago has also been great fun. Enjoy!

Graham Pitchfork
December 2023
Gloucestershire

CHAPTER ONE

FROM HUNTER TO SWIFT

NIGEL WALPOLE

One dreary evening in November 1955, after a few beers at 'Dirty Dicks', opposite Liverpool Street Station, we fledgling fighter pilots boarded a train for Harwich, then an ancient troopship to the Hook of Holland. There, colour-coded trains took us to north-west Germany, to join some twenty-five fighter/ground-attack squadrons of the Second Tactical Air Force. We would be flying one of the best day fighters of the time, the Hunter, and the fifteen-hour introduction to the earliest version at Chivenor was enough to whet my appetite for more.

Nigel Walpole.

My destination was RAF Oldenburg Lower Saxony once a Luftwaffe station with excellent facilities, now renovated and added to by the RAF Airfield Construction Branch. No. 124 Wing, comprising 14, 20 and 26 Vampire Squadrons took up residence in 1953, each squadron fully mobile with its own transport, and each partnered by an RAF Regiment squadron, equipped with 40-mm Bofors Gun. It was dark and pouring with rain from low cloud, typical of the North German Plain, when we four 'sprog' pilots entered the beautiful old Luftwaffe officers' club, now our officers' mess. There, we were welcomed by the mess manager, the legendary Herr Lach, an elegant, calm, and respectful professional, and one of the many attractive ladies he had chosen to serve us with our every need. Our bachelor quarters were luxurious and there was no food rationing here; drink, cigarettes, state-of-the-art electronics and tax-free cars were plentiful and very cheap.

When we arrived the three squadrons were converting from Sabres to Hunters, and I was destined for 26 (AC) Squadron, already fully equipped with sixteen brand-new Hunter F4s, ultimately to be manned by twenty-four pilots mostly 'first tourists' with very few hours on the Hunter at the OCU. With no two-seat Hunters we had access to the station flight's Vampire T11s for dual instruction and testing. Our boss was Squadron Leader J. A. G. Jackson; one of our flight commanders, Geoff Wilkinson, had flown with the USAF in Korea, and the other, John Cowley, was a talented ex-test pilot on the Hunter. They were supported by a sprinkling of second-tour qualified flying instructors (QFIs), instrument rating examiners (IREs) and pilot attack instructors (PAIs), but experience levels were generally low and we first tourists would be on a very steep learning curve.

Early in my tour our Hunters had Avon 113 and 115 engines which were likely to surge and flame out if handled roughly if they ingested hot gasses or the links between the rounds when the guns were fired. This was not good for combat against the MiGs we saw massing on the other side of the Inner German Border (IGB), but both technical faults were remedied during my tour, by 'fuel dipping' and the addition of 'Sabrina' blisters to catch the links.

After a ten-sortie orientation programme we concentrated on high-level battle formation, in pairs or sections of four, followed by a tail chase or simulated combat, with ranging and tracking recorded on film from cameras mounted on our gyro gunsights. Practice interceptions were carried out under close or broadcast control from the RAF's highly vulnerable Type 80/13 master radar stations, and a small number of mobile units. We could take our lightweight F4s up to 50,000 ft, to see well into East Germany on a clear day, where the MiGs were doing what we were doing, but lower down.

Recovery to Oldenburg, with the wing's short-range Hunters generating up to 100 sorties a day, in the perennial low cloud and poor visibility of north Germany, could be challenging. The only navigation aid we had in the cockpit was the distant measuring equipment, while our air traffic control depended solely on a cathode ray direction finder to help us; we had no precision approach aids. Moreover, we had the huge Delmenhorst mast rising above 1,000 ft at the start of our approach, and smaller masts either side of the final approach.

With our sky full of fighters, whatever we had planned invariably succumbed to simulated combat beginning with the 'players' seeking advantageous position and perhaps splitting their force before closing for an attack. Thereafter section leaders would orchestrate events as best they could. Wingmen might attempt to 'stick and report' but in the inevitable confusion

Hunters of 26 Squadron.

that would follow most would get separated from their leaders and it could become every man for himself.

Then there were invitations between fighter wings, usually on Saturday mornings to meet, say, over Dümmer See at 10:00 at 40,000 ft, which could snowball into a veritable circus of Hunters, Venoms (never try to out-turn a Venom below 40,000 ft!), Canadian and American Sabres and perhaps the odd Meteor. Of course, we all got there a little early and at our maximum heights, for a 'battle' which would range from high to very low level (and very low fuel states).

I'm sure the hierarchy knew about these unofficial activities, but perhaps with 'blind eyes', because they knew how much we learned from them, and in my time, I know of no accident that could be attributed to such sport. It might, however, have been different when RAF Jever Hunters took on 26's 'Bunny' House's 'four-ship' in mock combat. Fortuitously, Bunny's No. 4, Bob Snare, was 'channel hopping' and heard a ground-control intercept radar vector a Jever section on to them. He then spotted and continued to report the predators homing in on them from an estimated twelve miles in their six o'clock. In the 'combat' which followed he successfully evaded the nine 30-mm rounds fired at him accidentally. A guest night followed to mend fences between the two wings, but the less said about that the better.

We took our turn on 'Battle Flight', with guns loaded, and went to various alert states, often culminating in cockpit readiness and then 'scramble', never sure whether this was the real thing or yet another practice. On one such scramble Bob Snare competed with the USAF battle flight to help the pilot of a U-2 spy plane when its engine flamed out over East Germany. The Americans got there first, but by then the U-2 pilot had relit his engine – and everyone went home. On another Battle Flight scramble, it was Bob again who was sent to investigate an 'unknown' blip from the east, heading for the UK high and fast. This looked serious, but with Bob's Hunter carrying full ammunition and external wing tanks, he had difficulty closing on his quarry until they were over Amsterdam, still waiting for crucial orders. There he recognised a Russian Tu-104 'Camel', which frantic investigations on the ground finally revealed had been late in securing air traffic clearance for the civilian Camel's inaugural flight from Moscow to London.

In September 1956, 2ATAF acted as enemy bombers, to test the air defence of the United Kingdom, and for this 26 Squadron was detached to the Dutch base at Eindhoven, inter alia treating this as a tactical deployment. There, we lived under canvas; not for us the luxury of hotel accommodation and hire cars, often enjoyed by present-day airmen, but they would never see that evocative sight of some 200 contrails, in parallel lines heading for East Anglia, and those of fighter command's best climbing up to meet them.

We were day fighters, practised in night take-offs and landings solely to be ready for night deployments, so we had little to do on the airfield at Eindhoven when the sun went down. That is until we spotted an ancient mountain cannon mounted on a plinth in the middle of a small lake outside our host's officers' mess and thought it would look better outside ours at Oldenburg.

First, the junior pilot was sent to ensure that the lake was not electrified, as we had been told it was, and when he survived, we planned to move our trophy during the last night of the detachment. So it was that in the dead of night, shadowy, half-naked figures rolled the gun off its mounting into the lake, using ramps from our tactical equipment, along the bottom, up onto a convenient bank and into one of our lorries, for a quick get-away into a convenient RAF unit on the far side of the airfield. There, with the help of our engineering NCOs, we dismantled our prize, dispersing the bits throughout the convoy and returned safely to Oldenburg, where the gun was re-assembled, lovingly polished and mounted proudly outside our mess. We then discovered that a muzzle-loaded thunder flash could hurl standard billiard balls high into the sky in the direction of Oldenburg town. We never knew where they fell, and fortunately ran out of ammunition before any harm was done.

Meanwhile, jealous pilots from other stations were hatching elaborate plans to make our gun their own. RAF Jever's station commander leading a strong team, took over our guard room, officers' mess night guard and MT Section, using our heavy plant to complete their task before the alarm was raised. RAF Ahlhorn and RAF Wunstorf were thought to be next to try, but when the USAF at Bremerhaven were reported to have borrowed a heavy lift helicopter to take the exercise to a new level, the commander-in-chief 2ATAF, the redoubtable Air Marshal Sir Patrick Bandon, stepped in. The 'Abandoned Earl', a splendid commander of the old school, was not averse to a little harmless mischief himself, but an international incident and casualties had to be avoided, and he ordered the cannon to be returned to Eindhoven before anyone got hurt. It had all been great fun, well planned and executed with great determination, epitomising the spirit of the day.

Then it all changed. One afternoon in 1957 I had taken off from a very active Oldenburg in a Vampire for a sortie of aerobatics and low flying but returned to an almost deserted airfield. There was just one head visible in the control tower to give me landing instructions, and two lonely airmen waiting outside their hangar to put my aircraft to bed, but otherwise there was no sign of life. It transpired that we had all been given a massive pay rise, and all the bars on the base were open.

Hitherto we had all been poor. Very few of us had cars, and even fewer wives, but we were living happily in the relative luxury in our messes, with rock bottom prices in the bar, wonderful flying and good sport of all sorts and welfare buses to take us to Hamburg, Amsterdam, and other desirable attractions. When the snow fell, and we were threatened with clearing the runway by hand, we planned escape and evasion cum survival exercises to the Harz mountains or the Sauerland, making good use of our only officers' club in Germany at Winterberg.

Then we became rich, and many started to think about those tax-free cars and/or wives, and in due course mess life gave way to an increased demand for married quarters, if you were above the age of twenty-five, or civilian hirings if you were not. The RAF would never be quite the same again.

Later in the 1957 came another 'bombshell', when British defence minister Duncan Sandys persuaded his colleagues that surface-to-air missiles could render fighters obsolete, and do a better job at less cost. Accordingly, his axe fell heavily on the RAF's fighter/ground-attack force in Germany, all the Venom and most of the Hunter squadrons disbanding almost overnight.

I was lucky to find a slot on one of the two Swift Fighter Reconnaissance (FR) squadrons, No. 79 (F) Squadron at RAF Gütersloh, commanded by Squadron Leader Hugh Harrison, with John Gale and Peter Terry as his flight commanders supported by a sprinkling of QFIs, IREs and PAIs. Most pilots on the FR squadrons were at least second tourists, already experienced on jet fighters or the role.

Supermarine quickly modified the already tough and resilient Swift for its new role, adding reheat (afterburner) to the powerful Avon 114 engine to make it the first RAF aircraft to join the front line with reheat. Controls and cockpit ergonomics were optimised for low level, the wings had saw-tooth leading edges and massive flaps to double as speed brakes. A 220-gallon ventral tank, which could be dropped in flight, increased the range. The extended nose accommodated three Vinten F95 cameras in left, right and nose oblique stations and two of the 30-mm cannons were retained for self-defence or ground attack.

The first FR.5s arrived at RAF Geilenkirchen for II (AC) Squadron in February 1956, and at RAF Wunstorf for 79 (F) in June '56. Having flown the Hunter, my conversion to the FR.5 in the hands of Peter Terry, using a static cockpit simulator, was short but quite adequate. Peter took me out to the FR.5 to check my cockpit drills and start up, and then went to the runway threshold for a visual check that the reheat eyelids opened fully when I selected reheat, and with his permission off I went.

No. 79 Squadron pilots with Nigel Walpole back row fourth from the right.

There was nothing spectacular about the take-off, but the FR.5 accelerated very well in reheat and when I had 'cleaned up' it was easy and pleasant to fly. The rudder response was poor, but the ailerons and elevators were light and effective at low level within the normal operating speeds of 420–540 knots, and the aircraft was manageable up to a maximum of 600 knots 'clean'. Importantly, the FR.5 gave a very steady ride at low level and high speeds.

Criticism that the aircraft had insufficient range to be of use in NATO's Central Region was ill-founded. Its great value had to be seen in the context of its assigned NATO role and our proximity to the IGB. At low level, the FR.5's 'rule of thumb' radius of action was 190 nm at low level, assuming a reheat take-off, low-level cruise at 420 knots, five minutes at 480 knots and five minutes at 540 knots, leaving enough fuel for one overshoot at Gütersloh. This was enough to cover our pre-planned and ad hoc war targets west of Berlin; typically, one of mine was a route recce between Halberstadt and Osterwieck. Assuming no electronic countermeasures to conceal us, or aids on board to navigate within or above cloud, we would try to remain in contact with the ground, routing to avoid known SAM and AAA sites and wary of mobile defences, navigating by map reading, compass, and stopwatch only.

In the FR.5 we were confident that we could get to our targets quickly and safely, take one pass to gather the information visually, get the photographic cover and remain unscathed to give real-time information to the battle managers for their rapid reaction. With the likelihood of enemy-listening posts or radio jamming, we would pass our visual sightings back in short, perhaps coded in-flight reports direct to front-line units, strategically positioned ground-relay units or aircraft at height behind friendly lines and amplify our reports on landing.

We would choose the best of the FR.5's three strip-aperture cameras for the job, and four- or 12-inch focal lengths at cycle rates of four or eight frames a second, individual lights confirming which camera was in use. The F95 caused little short of a revolution in low-level airborne photography, offering imagery with acceptable clarity and overlap at unprecedented high speeds and low heights – typically 480 knots at 200 ft. On landing, film magazines would be rushed to a mobile field photographic unit (MFPU) or a reconnaissance intelligence centre (RIC) and the first negatives could be ready for viewing on light tables within ten minutes, with selected prints following in twenty minutes.

The claim that the FR.5 could not survive in combat can also be challenged. At low level and in the right hands, with judicious use of reheat and the variable, 'barn door' flap-cum-speed brakes, the aircraft turned very well at low level. Our solo display pilots, Brian Seaman, Roy Rimington, 'Harv' Harvie, Roger Pyrah and others, showed us how well our Swift could manoeuvre for terrain masking and in combat.

Although our recce task invariably took priority, FR.5 pilots were trained to use their cannon in evasive combat and against very high value ground targets – typically, a nuclear missile convoy. Our tactical mobility took us to our annual armament practice camp at RAF Sylt, those officers not flying re-rolled to drive one of our three-ton Magirus lorries. This was almost as thrilling as flying, with the German police escorting us through Hamburg at break-neck speed.

At Sylt, we practised air-to-air gunnery at 10,000 ft, and once Peter Terry had sorted out the harmonisation our scores were quite creditable. With the Swift a very stable gun platform, the same was true in air-to-ground gunnery which we rehearsed on overland ranges.

A Swift FR.5 low over Germany.

The overall capability of the FR.5 to carry out its role (albeit without an enemy) was well demonstrated in NATO's Royal Flush Tactical Reconnaissance Competition in 1958, when Dick Green and I, representing the Swift force, were the only competitors able to follow a winding road in a cloud-covered valley in Germany's Sauerland, to find a military convoy and help 2ATAF win the Gunther Trophy again that year.

Another great virtue of the Swift was its robust strength, proven first by test pilot David Morgan's epic flame out and wheels-up forced landing in open countryside in 1951. He careered over a hill and squeezed between a farmhouse and an empty brick-built toilet – which he destroyed with one wing, and the aircraft flew again in six months. No. 79 Squadron's Pat King provided further proof of the Swift's survivability when he flew his FR.5 through the tops of trees in the Sauerland, to horrific hammering, rending noises and associated warnings in the cockpit. Undaunted, he recovered his Swift back to base, where the underside of the fuselage and ventral tank were found to be severely damaged. Pat feared a one-sided conversation with a very senior person which followed, but apparently the interview soon developed into an amicable conversation between two rugby fanatics, and Pat completed his distinguished career as the air commodore director of Flight Safety.

It was difficult to keep the Swift serviceable, but neither squadron failed to meet their NATO generation requirements, and invariably their annual flying task. On 79 Squadron this was due largely to the old-style leadership of our engineering officer Mr Bob Jude, a man held in great reverence by officers, NCOs, and airmen alike. With the FR.5, his men faced many frustrations; there were many hydraulic and electrical problems and with wings that were filled with sealant as fuel tanks, leaks abounded. Much of the ancillary equipment was buried deep in the fuselage so many of these defects required the engine to be lifted out at an angle through the top of the fuselage.

When Gütersloh was about to close for three months for runway repairs, we were scheduled to fly all our aircraft out to Sylt as the contractors moved in, but with a few hours to go an engine change was found to be necessary. All but one engine mech were already on their way to Sylt but he, Mr Jude and all the pilots swarmed over the aircraft, got the engine changed and ground run with minutes to spare. Flying Officer Kiwi Graves was the luckless pilot who drew the short straw to flight test the aircraft en route to Sylt – but all went well, and we spent that night in the 'Copper Kettle', one of the many night spots on the island of Sylt which stayed open as long as we wanted.

So ended my first four year of gainful employment on the front line in Germany. I had learned a great deal about the air defence and armed reconnaissance of the day, and about myself, and my great good fortune continued with an exchange tour with the USAF, flying the formidable RF-101 'Voodoo' in its tactical reconnaissance role.

CHAPTER TWO

CANBERRA STRIKE IN THE 1960s

TOM EELES

During my last year at Sherborne School in the summer of 1959, the RAF section of the school's CCF, of which I was a member, had its summer camp at RAF Laarbruch in Germany. In addition to enjoying air experience flights in the station flight's Chipmunk and Meteor, I managed to get a flight in a Canberra B(I)8 of 16 Squadron. Thus, a seed was planted that was to grow to fruition sometime later, although I was unaware of this at the time. At the very end of my three years as a flight cadet at Cranwell in July 1963, two fellow cadets and I were offered a weekend at RAF Wildenrath in Germany, flying there with our instructors in Jet Provosts. I was much impressed by what I saw, the Canberra's role of low-level strike/attack looked exciting, there was plenty of like-minded male company living in the officers' mess, and even female company in the form of schoolteachers. There was cheap booze and petrol, no purchase tax to pay on a new car and the whole of Europe on the doorstep. It all looked to be a good deal and a much better prospect than sitting in the right-hand seat of a V-bomber managing the fuel. At the end of my advanced flying training on the Gnat in March 1964, I was delighted to find that I was the only one on my

Tom Eeles with his Canberra B(I)8.

course at RAF Valley to be posted to the Canberra, then even more delighted to find I had been posted to 16 Squadron at RAF Laarbruch in Germany at the end of my OCU course.

RAF Laarbruch was the most northerly of the four RAF 'clutch' airfields in Germany, situated right on the Dutch border, and constructed in the 1950s on the site of a temporary advanced grass airfield used in the latter stages of World War Two. Remnants of the Siegfried Line fortifications still existed on the airfield's western boundary. When I arrived in September 1964 it was home to 16 Squadron equipped with the Canberra B(I)8, 31 Squadron equipped with the Canberra PR.7 and 431 Maintenance Unit. When compared to RAF airfields in the UK, it covered a much larger area, with squadrons located at the corners of the airfield in dispersed sites. Aircraft were parked in the open on individual pan-shaped dispersals protected by earth banks. The 16 Squadron site was surrounded by pine trees; we were discouraged from walking from the hangar directly through the trees to our aircraft because of all the unexploded ordnance left over from the fierce fighting in the area in 1945. There was a ring road around the circumference of the airfield, not wide enough for two-way traffic as the contractor charged for a two-way road but built a single-way one and pocketed the proceeds in a Swiss bank account. There was everything you could wish for at Laarbruch, a heated open-air swimming pool, a local bank, a well-stocked NAAFI shop, a school, with the associated female schoolteachers, a bowling alley, numerous clubs, and an officers' mess with a vibrant social life. Petrol coupons were available enabling cheap travel throughout Germany, the Netherlands and further afield. Compared with life at some of the remoter bases in the UK, we were very lucky.

The Canberra B(I)8 was different from the earlier marks of the aircraft. The pilot was now positioned on his ejection seat at the rear of the pressure cabin, offset slightly to the left under a new fighter-type cockpit canopy that gave superb all-round visibility. Perversely this canopy could not be opened so access was still through the original entrance door, then climbing up into the cockpit. There was only one navigator position, ahead of and below the pilot and without an ejection seat, his only way to escape in emergency was through the access door with a manually deployed parachute. The navigator sat on a swivel chair at a small chart table or lay down prone in the nose for visual navigation at low level. The nose had additional glazing to improve his view. The B(I)8 was specifically designed for day/night low-level interdiction, so had more armour protection compared to previous marks. A gun pack fitted with four 20-mm cannons could be installed in the bomb bay, flares could be carried in the forward half of the bomb bay for night target illumination, and two 1,000-lb bombs could be carried on the wing-mounted pylons. In the nuclear-strike role, the gun pack was removed as the nuclear weapon took up the whole of the bomb bay.

At this time NATO's strategy was known as the 'Trip Wire Policy', whereby any Warsaw Pact incursion into NATO territory would be met immediately by a nuclear response, since NATO's conventional capability was far less than the Warsaw Pact's. There were four RAF Canberra strike squadrons based in Germany as part of the Second Tactical Air Force, assigned to NATO for tactical nuclear-strike duties in support of this strategy. Their nuclear weapon was supplied by the USA and remained under strict US custodianship until released by a 'dual key' arrangement requiring both UK and US government authority. Whether this would have been forthcoming in the fog and confusion of war was thankfully never put to the test. Weapon delivery was by the low-altitude bombing system (LABS), whereby the aircraft approached the target at low level and high speed, pitching up into a 'roll off the top' aerobatic manoeuvre, and lofting the bomb forward. There was an alternative mode of attack which involved direct overflight of the target and pitching up into the same manoeuvre, but with the bomb being released after the aircraft had passed through the vertical. Each Germany-based squadron maintained two loaded aircraft on quick reaction alert (QRA) at fifteen minutes' readiness to get airborne 365 days a year. These were positioned in a guarded compound close to the runway, guards being provided by UK and US personnel.

The day-to-day flying consisted mainly of obtaining, and maintaining, currency in the nuclear-strike role. This involved single aircraft low-level navigation sorties throughout the Federal Republic of Germany, the Netherlands, Belgium and occasionally France, combined with LABS bombing on the RAF's weapons range at Nordhorn, in the north-west close to the Dutch border. A degree of caution needed to be exercised the further east your low-level route took you, as there was an Air Defence Identification Zone (ADIZ) running along the Inner German Border. Penetration of this was prohibited as it might result in you being shot down by the Warsaw Pact or your own side. There was another area to be careful of, running alongside the ADIZ, called the Buffer Zone. I recall that you could fly in this provided you were not heading east. It was mandatory to keep a listening watch on the guard frequency, and if the call of 'Brass Monkeys' was heard, all aircraft in the Buffer Zone had immediately to turn west. The proximity of a heavily armed enemy

Two Canberra B(I)8s of 16 Squadron airborne from Laarbruch.

was very real. If the weather was poor in Germany, high-low-high sortie profiles were flown to the UK, again using UK weapons ranges such as Wainfleet and Tain.

There were significant limitations placed by HQ RAF Germany's Air Staff on the conduct of LABS bombing because of some frightening early experiences during recovery from the manoeuvre in cloud and at night. Essentially, LABS bombing was prohibited if there was any likelihood of the aircraft being in cloud during recovery, and there was no LABS bombing permitted at night. Clear conditions around 4,000 ft, the manoeuvre recovery height, were rare in Germany in the winter months, so to maintain our currency in this challenging activity we would deploy regularly to sunnier climes in the Mediterranean. The RAF maintained airfields at Gibraltar, Malta, Libya and Cyprus, with weapons ranges at Tarhuna and El Adem in Libya, Filfla in Malta and Episkopi and Larnaca in Cyprus. We visited all these places regularly, a very enjoyable experience. Flying at low level over the desert in Libya, where there seemed to be no minimum height at which you could fly, was fascinating. There was still much debris left over from the great battles of World War Two, including armoured vehicles, fortified sites and of course the legendary remains of the B-24 Liberator 'Lady Be Good'. When operating from Idris, the airfield south of Tripoli in western Libya, Tarhuna range was so close that you could get airborne, deliver four 25-lb practice bombs by LABS then recover to Idris, load four more bombs whilst keeping the engines running, and deliver them, all within one hour. These trips to the Mediterranean facilities also generated a busy exchange market in goods unobtainable both in Germany and in the Mediterranean. The Canberra's capacious bomb bay and rear hatch gave plenty of storage space. Asparagus, strawberries, and fresh milk found ready recipients in Idris and El Adem, in exchange for sheep skins and dates. Casks of sherry were available in Gibraltar, wine, brandy, oranges, and lemons in Cyprus. It was considered very poor form to arrive at the remote outpost of RAF El Adem without something for the locals.

The fact that we were the tip of NATO's nuclear spear in the event of the Cold War turning hot did not seem to worry us unduly. Given that most of us on our first tours were under twenty-five, we considered ourselves immortal. Whilst it was a bit worrying to see that our target folder maps did not show any route home from our war targets, we carried on enjoying life to the full, assuming the unthinkable would never happen. Of course, there were the inevitable exercises, some generated by the station called Minivals, some by HQ RAF Germany, Maxevals, and finally the major NATO 'no notice' exercise, Taceval, which assessed the whole station's ability to meet its war tasks. Failure to reach the required standard on Taceval would have serious consequences for those in command positions on the squadrons and on the station. An increase in the alert state would be signalled by the loud mournful booming of the station hooter, audible everywhere, usually at some ungodly hour of the morning. After dressing rapidly and driving as fast as possible to the squadron, you would find out if it was just an exercise, or for real. The crews in QRA would have manned their aircraft, but power would not be applied if it was just another exercise. If you found a whole lot of stern-looking officers from other NATO air forces with clipboards in the squadron building you knew it was Taceval. The squadron would be tasked to generate as many aircraft in the nuclear-strike role as quickly as possible. This was a time-consuming, combined British and American, procedure, as the Canberra had to be jacked up to raise the tail sufficiently for the nuclear weapon to be loaded into the bomb bay, then the aircraft was surrounded by a canvas screen behind which the American armourers did something that the British were not privy to. It was rare to load the real weapon, on most exercises the concrete version, known as the shape, was loaded. There was always a time limit by which all this had to be achieved. Then it was usual to have a flying phase. The aircraft would have the shapes downloaded and practice bombs would be substituted. A launch procedure would be practised, culminating in a mass launch and a sortie culminating in first-run attacks through Nordhorn range. One lucky crew might find it had been selected to hurl one of the enormous concrete shapes at the target. Another exercise input might involve the selective release procedure, simulating a situation where the enemy had made a breakthrough and the rapid delivery of a tactical nuclear weapon into the land battle being the only way to stop them. A device known as the weapon response simulator could be fitted to the aircraft, which allowed the crew to go through the somewhat complicated switching procedure to release a nuclear weapon, which was simulated by the release of a small 25-lb practice bomb.

A Canberra B(I)8 of 16 Squadron escorts four USAF F-105s over Laarbruch.

Winter survival course at Bad Kohlgrub. Tom Eeles is in the front row, third from left.

Another most enjoyable feature of our life in Germany was the annual NATO squadron exchange programme, whereby the squadron would deploy to another NATO nation's air base to fly alongside the host nation's air force for usually a week, with a return event being held back at Laarbruch. The accompanying photo shows four F-105 Thunderchiefs in formation with one of our Canberras over Laarbruch, with the QRA compound clearly visible in the background. The Thunderchiefs came from the USAF base at Spangdahlem, in southern Germany – their role was identical to ours and they were equipped with the

same nuclear weapon. Naturally there was a vigorous social programme associated with these squadron exchanges, introducing the host nation's culture to the visitors, and vice versa, so flying often started late the next day. A slightly less enjoyable experience was the mandatory requirement to attend the RAF Winter Survival Course at Bad Kohlgrub in southern Germany, where participants practised survival, escape and evasion, followed by interrogation, in winter conditions. The bonus was that you learnt to ski for free.

The conventional weapon capability of the Canberra was exercised infrequently, as this role would have been used solely in support of UK national activities. It was not part of the RAF's NATO commitment. Fortunately, our squadron commander, who had previously served on Hunter squadrons, encouraged tactical formation flying as much as possible in pairs or fours during our normal daily flying to maintain this important skill. The squadron would be tasked with carrying out one or two three-week-long armament practice camps (APC) each year, usually overseas at Akrotiri in Cyprus or Luqa in Malta. QRA had to be maintained during these APCs, but the remaining aircraft would have a role change, involving installing the gun pack in the rear half of the bomb bay, fitting specially modified doors to go round it, and fitting the flare beams in the front of the bomb bay. These APCs were immensely popular and were much looked forward to. We would practise shallow dive-bombing in thirty-degree dives, aiming with the rather primitive gun sight, and air-to-ground firing of the 20-mm cannon in ten-degree dives. At night we would try to illuminate the target using the flares, but they were all rather old and unreliable, so it really was rather a hit or miss business. I well recall being nominated as the range safety officer (RSO) one day at Larnaca range in Cyprus, now the site of a civil airport. This involved setting off from Akrotiri at

No. 16 Squadron on detachment to Kuantan, Malaysia.

05:00 in a three-ton lorry with the rest of the ground party, negotiating our way through various Greek and Turkish communities and their roadblocks on our way to the range. Once there, my main role as RSO, apart from monitoring the aircraft's activities, was to attempt to deter the many Turkish youths from entering the danger area whilst the aircraft were firing. Their intention was to collect the brass shell cases which fell from the firing aircraft, which were then melted down and turned into pots, pans, and tourist trinkets. I was equipped with a Very pistol and many red flares to do this, the idea being that if they intruded, I was to fire red flares at them. They all understood that they were not to enter the danger area while aircraft were using the range, but inevitably one hot-headed youth would always break ranks and rush onto the range to be the first to collect the precious cases. There was a danger of serious injury from the heavy shell cases falling on them, with the RAF now being liable for compensation. I recall running out of red flares and having to close the range to restore order. In January 1965, 16 Squadron was stood down from its QRA commitment and nine aircraft were sent in the conventional role to reinforce the Far East Air Force during the confrontation with Indonesia. They were housed at a bare base airfield in Malaya near the town of Kuantan, living under canvas, and remained there until June. I was lucky enough to persuade my squadron commander to let me join this wonderful detachment despite my lowly status as a junior pilot, but that is another story worthy of a separate chapter.

In conclusion, it is interesting to surmise as to how effective we might have been in those days. Whilst the fact that NATO fielded many nuclear-armed aircraft on QRA in Germany doubtless had a deterrent effect on the Warsaw Pact's leadership, if the Cold War had turned hot our capabilities were limited in the nuclear-strike role. The lack of any self-defence equipment in the Canberra meant we were vulnerable to fighters, surface-to-air missiles, and anti-aircraft artillery, of which the Warsaw Pact had plenty. There was a small radar known as Orange Putter that could be fitted in the Canberra's tail to give warning of approaching aircraft, but none were fitted to our aircraft. Our training did not include weapon delivery at night or in bad weather, and the opposition must have been aware of that. The QRA compounds were completely 'soft' and vulnerable to air or special forces attack. We relied solely on flying low to avoid being shot down, and no evasive manoeuvres were officially taught or practised. In the conventional role we were no better armed than a World War Two fighter such as a Typhoon, the only advantages being having a longer range, a slightly higher speed, more ammunition, and a rather dubious night capability. Nevertheless, I much enjoyed my time as a Cold War Boy overseas in Germany despite the above deficiencies. The opportunity to fly all over western Europe, the Middle East and, unexpectedly, the Far East was a wonderful start to my military flying career. However, I moved on after nearly two years to loan service with the Fleet Air Arm flying the far more up-to-date and capable Buccaneer.

CHAPTER THREE

WELCOME TO THE GOLDSTARS

GRAHAM PITCHFORK

With the Canberra PR course at Bassingbourn nearly complete, my pilot, Terry Close, and I learned that we had been posted to 81 Squadron in Singapore, but with just two weeks to go, 31 Squadron lost a crew in a tragic flying accident. NATO was deemed to have a higher priority than the Far East, so Terry and I found ourselves cancelling our tropical kit and heading for Laarbruch to join the Goldstars in April 1962.

The squadron was equipped with the Canberra PR.7, which had much greater range than the PR.3s we had trained on at Bassingbourn. Half the bomb bay held a fuel tank with a photo-flash crate for night photography occupying the rear half when it was not full of aircrew luggage and spoils from trips to far-flung parts of the world.

Graham Pitchfork.

Pilot officers were virtually unheard of in the PR world and the first comment made to me by the CO (Wing Commander Charles Dalziel) on my first morning on the squadron was less than encouraging. "Shouldn't you be up the road with that bomber lot [16 Squadron] full of first tourists?" he said with a twinkle in his eye. He then informed me that it was a squadron tradition to fly newcomers on their first day.

My flying kit had not arrived, neither had my pilot and the weather was filthy. No problem as far as the boss was concerned. I was kitted out, introduced to the pilot, Squadron Leader Ramsey Brown, and piled into the back of a PR.7. His navigator was a rather older gentleman who had once served in the Indian Cavalry, and who pre-fixed every noun with "wizard". I don't recall being offered a map.

The weather was poor, so 'Wizzo' occupied the rumble seat, which left me stuck in the back, couldn't see a thing and hadn't a clue what was going on. It soon became apparent that this situation had also afflicted one other member of the crew. I had vague sightings of the outside world from the back seat and was somewhat confused by the ramblings from the cockpit as Wizzo thrashed about with his large map. Understanding him was not helped by his affliction of a heavy lisp. "We will soon cwoss a woad over a wivver...oh dear... it's a wailway over a woad, never mind."

After other similar mutterings, Ramsay's frustration was beginning to break through his calm demeanour and, thankfully, he decided to abandon the sortie and return to Laarbruch. Easier said than done since knowledge of our position was scant and it was extremely murky, but he managed it. My logbook tells me that we were airborne in WT533 for almost two hours and the remarks column records 'low-level nav demo' – a charitable description without doubt.

My initiation to the squadron was far from complete. The self-appointed senior flying officer and self-appointed chairman of the officers' mess bachelor's club was one Tony Lovett, who I suspect had a low opinion of Cranwell graduates. I was programmed to fly with him, and we were given the usual five targets in the BZ (British Zone). Tony added a sixth, which appeared to be a few miles off the coast at Cuxhaven on a low-lying island. Having found the first five, we left the initial point (IP) on the coast and headed west on a timed run to photograph Tony's target. I looked ahead and saw nothing, and panic began to set in. On time, there was no sign of an island, or a target and it was clear I had missed it. We returned to Laarbruch in silence, and I had to admit to failure during the debrief. Later that evening I heard Tony's voice in the bar explaining that he had just "fixed that new nav" by selecting a sandbank knowing that the tide was in, and it would be underwater.

A few days later, I was programmed to fly a check ride with the CO. It was another five targets in the BZ culminating with a 'military installation' near Bremen. All went well until this final target. We set off from the IP but at the appointed time there was nothing that looked remotely like a military target. On landing I noticed that the other crews took an unusual interest in our sortie, and I was to learn later that this was the CO's party piece for all new navigators. When the film arrived, the CO took me through it until the fifth target should have appeared. He had taken some photographs, and I could just about make out some old and derelict buildings. I enquired if that was the target, "Yes my boy," came the reply, "they are the remains of the camp where Herr Hitler incarcerated me for two years after I had been shot down". He then gave me the most important piece of advice for those in the PR business. "Always take a photograph at the elapsed time, the photo interpreter may find what he is looking for even if you don't see it." And with that he took me off to the bar.

Although the squadron was tasked primarily as a low-level tactical reconnaissance squadron using the oblique F95 camera, there were also other interesting tasks. I never discovered the reason for some of the high-level photographs taken over France. We also occasionally flew at medium level, and with a very long focal-length oblique camera flew along the Inner German Border looking into Eastern Germany. This exercise warranted a Javelin escort.

There were several regular features during the year including no-notice station exercises and Tacevals, national and NATO reconnaissance competitions, regular flights to NATO airfields and lone-ranger flights to the Mediterranean and beyond.

During an alert exercise, there was a big advantage being on a PR squadron. Pre-planned sorties were launched immediately to carry out a line search to find the enemy's advance. Each of the three PR squadrons in Germany launched two aircraft as soon as possible – our 16 Squadron bomber chums sat in the crew room for a few hours waiting for our results. There was, therefore, a great incentive to get to the squadron immediately since the first two pilots and two navigators to appear were crewed together and sent off. It became a competition to see who could arrive at the squadron first. Around Laarbruch there was a five-mile one-way ring road. The squadrons were dispersed in each corner of the standard NATO airfield and No. 31 was the furthest away from the domestic site. However, by going the wrong way round the ring road, those of us in the officers' mess had the shortest route. Since all alerts happened at the dead of night, no problem. It took the married chaps ages to discover our secret.

Two aircraft were always parked overnight in a dispersal near to the hangar. The ground crew rushed to prepare them as the first pilot grabbed his kit and raced for the aircraft. The nav was given a pre-planned route folder before heading to the aircraft that now had the engines running. Take-off was often achieved in less than thirty minutes from the alert sounding. Photographs were taken of the line search and then a high-speed return flight to be met by the photographers who soon delivered the film magazines in the mobile field photographic unit (ours was 3 MFPU) and they had the film processed for interpretation within a few minutes.

The squadron had a variation of this rapid capability devised by our new boss, Wing Commander Peter Scott, designed to impress VIPs. As the fittest nav, and one of few who could run, I was selected for this demonstration. On arrival, the VIP was met by the boss and brought to a planning table where I had a map carefully folded so that it only just crossed the River Rhine. Our guest was invited to select a target, I drew a few lines and then sprinted to an aircraft parked by the hangar. Engines were fired up as I approached and off we roared. The VIP then toured the squadron.

I never had time to strap in, passed a heading to the pilot and at full speed we headed for the target. We took two photos, one with each F95 camera and raced back making lots of noise at very low level over the hangar to herald our return. I fixed a dayglo sticker in the nose to indicate the priority camera. We taxied up to the MFPU, where the VIP and reception committee were waiting, engines were cut, the photographers had the magazines off, and I strolled to debrief our guest. A few minutes later he had a photograph of the target presented to him. One up on the bomber boys!

Overseas trips were a regular feature of life on a Germany-based squadron. In December 1963, Terry Close and I headed for Gibraltar. One hour away, we received a radio call giving us a mandatory diversion to Oran since the crosswinds at Gib were too severe to attempt a landing. This created something of a stir in the aircraft's map department, resulting in having to make use of an en-route chart to determine where our new destination was. Arriving overhead Oran at high level, I spotted three airfields close to the town. One was clearly a large commercial airport, but it needed a low-level sweep to identify the correct one. With one obviously abandoned, the problem was solved.

High-priced help for the visit of the Minister of State (Air), Hugh Fraser MP.

Even before landing, we became aware that there was an air of disorganisation about the place. On our downwind leg, a French air force P-47 Thunderbolt of World War Two vintage ranged alongside and proceeded to do a barrel roll around us. Matters did not improve once we landed, and after a tour of the airfield's outer extremities, we were directed to a dispersal. It took time to organise refuelling, the operations room showed little interest in us when we reported, and we had difficulty contacting Gibraltar to alert them to our safe arrival. The overnight accommodation we were offered was beyond description and, despite the heat, Terry made it clear he was going to sleep in full flying kit to avoid becoming a host to the many and various bugs whose home he was about to disturb.

When we headed for the officers' mess, our luck began to change. We had barely had time to buy a drink when we were approached by a French officer in flying kit. Incredibly, we had met him a few weeks earlier at Metz during a NATO turn-round exercise. He explained that the French were about to pull out of the base, hence the spartan conditions, and he had flown down for one last visit having been stationed there in earlier times. The rest of the evening took on a different character as we were invited to join a French air force 'Happy Hour'. Much later, on retiring to our room, the company of our bedfellows seemed less of an ordeal.

The following morning, a Saturday, we headed for Gib at low level to be informed that the wind was less severe but still tricky. The runway in use was the westerly one, which always created the most turbulence when the wind was south-westerly. The next forty minutes could best be described as 'exciting' as Terry made numerous attempts to land. Finally, we made it to be met by the OC Operations Wing who informed us that, had we not succeeded on our last attempt, he was about to divert us...to Lisbon!

Competitions were a major annual feature for a PR squadron in Germany. The first was the RAF's Sassoon Trophy competed for by the three Canberra squadrons (Nos. 17, 31 and 80) and two Hunter fighter reconnaissance squadrons (Nos. 2 and 4) based at Gütersloh. We carried out weeks of training before five, and a reserve crew, were selected for the competition. In our first year, Terry Close and I were included in the team. As the junior boy under the spotlight I was, quite naturally, very nervous.

Crews took off at one-hour intervals throughout the day. Each aircraft in turn was positioned on the taxiway just outside the squadron planning room some fifty yards away. The crew prepared the aircraft in advance and reached it by jumping out of the crew room window.

The umpire issued the task of three targets with an entry and exit point. After a very close assessment of the geographic features around the target, the direction of approach was decided for the port oblique. The nav selected the initial point and drew the target run as the pilot prepared a route map. The entry and exit points had only to be recognisable on the photograph, but the targets had to be perfect for the photographic interpreter (PI) to prepare his post-flight report. Thirty minutes was allowed from issue of the task to engines start.

We were third off and had some tricky targets in the 4ATAF region. The first, a radio mast, was cleverly positioned a few hundred yards from a radar site but the adage of taking a photo of the exact grid reference, and not being suckered in by a more obvious 'target', worked. The second was a bridge buried in trees in the Ardennes and the last was a line search for a POL (petrol, oil, lubricants) site. Panic set in as we approached the end of the line having seen nothing. At the last minute I spotted an army fuel bowser just off the road. We took a photo in hope. On landing we discovered it was the target!

As the engines were shut down, the magazines were rushed to the MFPU for processing as we made our visual reports, which were then amplified by the arrival of the photograph negatives, which the PI studied with his stereo glasses. His report had to be handed to the umpire within thirty minutes. The team had done well, and the squadron were declared the winners for the 1963 event.

The NATO competition, Royal Flush, was a much bigger affair with all the Central and Northern Region countries represented. In the long-range category, two of the three RAF Germany squadrons competed, one by day the other by night.

In 1964, No. 31 was the RAF's entry for the night phase. Three crews were selected, each flying three sorties on consecutive nights. Our main opponents were the USAF RB 66s flying out of Alconbury, a specialist night-reconnaissance unit with a four-man crew, including a radar operator. Our two-man day outfit was up against it. To enhance our chances, a second nav was used, and Flight Lieutenant Jim Straughan, an ex-night fighter nav, joined us and he operated the Green Satin to give the pilot accurate ground speed and drift, and he also made the settings on the F97 camera panel. It was my job to do the route navigation from the nose, and, more importantly, the IP to target run on what proved to be very dark nights.

Halfway through the night work-up my pilot Terry Close was medically grounded. Ramsey Brown's nav was not enjoying much success map reading at night so he inherited Jim and I to form a crew.

The work-up was very unsociable with three long night sorties a week for eight weeks – and it was springtime so some very late nights. We practised on just about every weapons range in Western Europe to work out routes, IPs etc.

Accurate flying was essential with the ground speed nailed at 300 knots, the height of exactly 1,000 feet above ground at the target and tracking directly over the unseen target. Ramsey Brown was a big man with huge hands, but his flying was immaculate and very precise. We soon became a very effective three-man team.

Just before 2300 hours on 12 May 1964 we took off on our first sortie and headed for the Stanford training area in Norfolk for the first target, with the second on a Belgian range and the final one at the familiar Nordhorn range east of the Rhine. The flashes went off at two-second intervals with the target, hopefully, covered by the middle of the stick. I was able to confirm we had the targets.

The second night was to three more ranges, all of which we had visited in the work-up. At the end of this night, we got the whisper that the squadron was in the lead, and we were the top crew. No pressure!

On the third night we were tasked to about the only range we had never seen – in Denmark. The second was at Vliehors in the Frisian Islands and then to Suippes in northern France. The Danish range was the toughest one we faced. There was very little to see, and I had to rely on Ramsey's accurate flying and Jim's constant monitoring of the Doppler. In the glow of the flashes I saw some tanks, which was very encouraging. We were familiar with Vliehors and all went well. Suippes was a very different matter. The range was big so the IP to target runs were long with few features recognisable at night. On an earlier training sortie, we had found a small cluster of lights around a compound in the range so chose that as the IP. We arrived after midnight– no lights! A hasty decision to use a nearby river with a prominent bridge was made and we headed into the black. I released the flashes and there below us were three Harvard aircraft simulating missiles.

The results came out the following day. The squadron was beaten by the narrowest margin (it would not have done to beat a specialist night squadron) but the Brown, Pitchfork, Straughan team had carried off the crew award. We celebrated in fine style before re-discovering what flying in daylight was all about.

By the end of Royal Flush, I had been on the squadron just over two years, and having lost Terry Close permanently, the boss gave me two options. Early tourex, almost certainly

Graham Pitchfork and crew return from a night Royal Flush sortie.

to the V-Force, or extend my tour and be his nav for the next twelve months. He sent me from his office to think about it. One nano-second later I knocked on his door and told him I had thought about it and would stay. What a great decision that turned out to be.

Towards the end of a tour, each crew flew an 'Extended Southern Ranger' to Aden and Nairobi with some lucky ones continuing to Salisbury. Our turn came at the end of January 1965. After days preparing maps and flight plans, the boss approached me the night before we left to tell me that he had invited OC Operations Wing to come with us. He hoped it "would not cramp my style being with two wing commanders".

The boss was a keen skier, and as we loaded the aircraft before departure, he produced a pair of skis. On the return through Cyprus, which he had decided should be over a weekend to spend time with one of his big mates, Akrotiri's OC Strike Wing, he planned to ski on Mount Troodos. Unfortunately, the skis did not fit in the flare bay so, after I had climbed into the back, he managed to slide them into the cockpit and along the nav couch in the nose.

We headed for Cyprus for the night and the first thing to appear as the ground crew opened the door were the skis followed by two wing commanders and a tired and cold flight lieutenant. There were a few strange looks. Next morning, we loaded up the luggage and the skis appeared. "But, boss, would it not be a good idea to leave them here and get them on our return?" Back came the answer: "No, we might go u/s en route and not come back here." So off to Bahrain on the trickiest leg for the nav using only Green Satin and a GPI. I had prepared topographical maps for the route but there was no way I could reach the nose. I soon discovered that OC Ops, a former Javelin squadron commander, was not at his best lying in the nose map reading over the Caucasus from 40,000 feet.

The charade with the skis was even funnier in Bahrain and became hilarious on arrival at Khormaksar where the ground crew clearly thought we were crazy. So on to Nairobi and more astonished ground crew. On the return flight to Cyprus the boss got his two days of skiing. The aircraft went mysteriously unserviceable in Malta, so it was another night in the company of two brilliant wing commanders who were probably enjoying their last squadron days after long flying careers.

The winning crew. From left to right: Graham Pitchfork, Squadron Leader Ramsey Brown and Jim Straughan.

Aircrew of 31 Squadron mount a Canberra PR.7.

So, a fantastic tour with some great people came to an end after three years. I was lucky to fly with some excellent pilots, help win the Sassoon Trophy and be in the winning crew on night Royal Flush. I also shared duties and a social life with outstanding people, and we were lucky with our COs, Charles Dalziel and Peter Scott. Warrant Officer Wildman and his professional and loyal bunch of ground crew could not have supported us better.

Life on a PR squadron in Germany for a first-tour bachelor could hardly get any better. Almost 1,000 hours of flying in a fascinating and very varied role, cheap booze and petrol, lots of sport and a great social life with an officers' mess full of young bachelors and some very attractive young ladies. What a start for any young officer's RAF flying career.

It is always said that the two squadrons that mean the most to you are your first and the one you command. I could not agree more, and the 31 Squadron badge and the photographs on the wall of my den at home are reminders of a wonderful first tour on a magnificent squadron.

CHAPTER FOUR

HARRIER FIELD OPERATIONS IN GERMANY

IAN STEWART

SETTING THE SCENE

Germany was the front line of the Cold War in Europe in the 1970s and 1980s. NATO in West Germany looked across the Inner German Border to the Warsaw Pact forces in East Germany and beyond. Our RAF Harriers formed a small but unique part of NATO's air power, and I was privileged to have flown three tours (and completed a ground tour) in Germany over those two decades.

Ian Stewart as OC 1 Squadron.

For young, spirited aviators the flat plains of Germany hardly summoned up the excitement of earlier flying adventures in the Far and Middle East. But in the 1970s a posting to Germany was very popular amongst the fast-jet community; the country offered an expansive and interesting airspace, well-equipped bases, an excellent low-flying system, a welcoming and friendly local population who largely supported the need to train realistically, and the opportunity for close cooperation with Allied squadrons. The flying was challenging, honed by regular, demanding tactical evaluations; airborne hours were plentiful, the units were well supported by HQ RAFG, and defence funding at the height of the Cold War received relatively high priority. The assessed threat from the forces across the border was taken very seriously, motivation amongst the air and ground crews was buoyant, and a genuine 'can-do', positive spirit permeated the whole command.

The Harrier Force possessed two unique and important characteristics: from 1977 to 1992 it operated from the most forward RAF operational base in Europe, just eleven minutes flying time from the IGB and sixteen minutes from the nearest Flogger fighter base; and the Harrier possessed the ability to operate from dispersed field sites even further east, courtesy of vectored thrust, allowing short take-offs and vertical landings. These two defining factors are the focus of this short article on the Harrier in Germany.

I arrived in the country in October 1974, initially at RAF Wildenrath, just before midnight after a long, wet and windy journey in convoy with another young pilot, both of us driving fairly clapped-out old motors and excited at the prospect of the forthcoming tour. We were met at the somewhat forbidding gatehouse by a stern, no-nonsense flight sergeant policeman who told us to report immediately to our squadron. Dog tired and used to japes from time on the Hunter and the Harrier conversion unit, we foolishly trekked off to our allocated rooms in the old wooden shacks and hit the sack. Not for long. The same grumpy copper pitched up within minutes and read the riot act. Within minutes we were hat on in front of the 'squadron commander's' desk being dressed down by what appeared to be a very young, and clearly highly inebriated chap in wing commander kit having a small rant. He explained that all hell had let loose on the eastern front, and we were required for immediate duty. Yes, another jape, but the start of a wonderfully satisfying and sociable time in the theatre.

FORWARD BASING – UPFRONT AND PERSONAL

The Harrier, affectionately known by some as the 'bona jet' and annoyingly dubbed 'jump jet' by the press, was unconventional, quirky, and a joy to fly. It had lots of downsides – it was small, single-engine, carried a comparatively meagre payload for ground attack, had limited range (especially in the early Mark 1 days), could bite in the transition from wing- to jet-borne flight, and the cockpit was cramped and an ergonomic mess in the early versions. But vectored thrust from the four swivelling nozzles on the airframe transformed this baby ground-attack fighter into a versatile delight for those who mastered the skills of transition and jet-borne manoeuvring. Short take-offs were the norm, utilising a degree of vectored thrust dictated by the nozzle angle, and there were many options for landing depending on weight, from vertical, to rolling vertical, to short and even conventional, the least preferred but practised in case vectored thrust was unavailable.

Most daily Harrier flying training in Germany was, as for the smaller sister squadron and the conversion unit back in the UK, from fixed bases. RAF Wildenrath was set well back from the IGB. From there, for any eastward flights into the designated low-flying areas, we had first to weave through the heavily built-up areas – and, in the early 70s, often through the murk and industrial smog emanating from the Ruhr factories – between Düsseldorf, Cologne and Bonn, before commencing any meaningful training. On one of those transits at 1,000 ft, a rogue bird, always a potential hazard if swallowed by the engine via the 'elephant ear' intakes, crashed into the armoured front windscreen of my trusty steed, then into the canopy, which largely disintegrated. The wind noise was deafening. Forward vision was obscured by broken glass and Perspex, and my comfortable if cramped 'office' was invaded by hydraulic fluid leaking from the windscreen wiper (yes, we had a wiper for clearing the screen in the hover!), a long strand of miniature detonating cord hanging around my neck and helmet, and lots of smelly blood and guts (the bird's, not mine). The noise drowned out any radio comms from air traffic until a young female apprentice, with a piercingly high voice, linked me up with a shepherd aircraft to lead me back to base. Well done, SATCO. And I learned that, just because you cannot hear transmissions, you do not have to shout back.

We moved in early April 1977 to Gütersloh, fifty nautical miles east. We took the base over from the Lightnings, and on arriving on the dead side of the airfield at the appointed time we observed their rather exciting farewell pass across the pan in front of air traffic. (We heard the sortie became even more interesting later when one of the departing formations decided to 'celebrate' his arrival back in the UK with a few impromptu fly-pasts.) This move eastwards to Gütersloh seriously cut unproductive transit time, made compelling operational sense, and had the added advantage of moving us deep into the 1 British (BR) Corps area alongside many of the troops we supported.

Once there, one of my first instructions from the boss was to spend two weeks on a major ground exercise with corps troops up north, on the Bergen-Hohne training area, living in the back of an army personnel carrier, to become acquainted with the life of the troops we supported. The days were very long, the conditions basic, and a world away from the relative comfort on base. I learnt three key lessons: never sit right at the back of the carrier or the hot water tank in the rear door (heated by the exhaust) will spill and scold at every bump in the terrain; sleep when you can, not when you want, as the 'war' never stopped; and listen very carefully to the experienced non-commissioned officers who generally call the shots in the field and know their jobs backwards. Though I would rather

Harrier GR3 lands at a dispersed site.

have been 250 ft above the fray in a warm cockpit, the experience was character building and left a lasting respect for the boots on the ground, and their genuine appreciation of getting dedicated air support both from recce over the next hill and in providing welcome tactical airborne support.

Gütersloh was a popular, vibrant station, with two eighteen-aircraft Harrier squadrons, two support helicopter squadrons (Puma and Chinook), a regiment air defence unit, and more. It was the trooping airhead for 1BR Corps, and played a major role in enabling the British, French and US access to Berlin down the three established air corridors across

East Germany. Our main 'customer', 1BR Corps, was headquartered nearby at Bielefeld, though we also supported Belgian, Dutch, and German ground troops. The Corps Air Support Operations Centres tasked the Harriers regularly in support of their troops. Daily flying was demanding but great fun, whether working in the close-air-support role, the deeper battlefield air interdiction, reconnaissance, or practising air combat in the local dedicated airspace above the airfield. The local air-to-ground range was a short hop away at Nordhorn, though at least once a year we would deploy for two weeks on armament practice camp to Decimomannu base in Sardinia, to use the excellent facilities at Capo Frasca range, and later the medium-level air-combat installation for dissimilar air-combat training.

Most days on base we planned the sorties, drew up the maps, briefed, then headed east, north, or south, into the low-flying areas to practise attack profiles and support the various ground units, either with forward air control or on pre-tasked missions, generally as pairs, often as four-aircraft formations, frequently against 'bounced' opposition. Add the occasional night flying, the regular supervisory check sorties, weapons competitions like the Salmond Trophy, exercises, air tests, squadron exchanges, and land-aways to other NATO bases, and most pilots clocked twenty to thirty hours of high value airborne time each month. Though the hours may sound puny compared with those flown by many other fleets, most sorties in the GR3 era were under fifty minutes duration, prefaced by two to three hours preparation. Once the much more capable Harrier Mk 2 arrived, in the late 1980s, with extended range and payload, sortie durations increased. At the end of each sortie, we would loiter around the airfield and hone our VSTOL skills on the vertical landing pads and the short strips on base before taxying back to our hardened aircraft shelters. Air traffic on the geographically small, very compact station was often hugely busy, with multiple aircraft in the circuit – fortunately we were blessed with some highly talented controllers. Then back for coffee, and the all-important frank post-flight debriefs, to analyse the sortie in detail and learn how to do it all better in future.

On our return from one bounced four-ship sortie we were taxying in towards the hardened aircraft shelters on the south side of the airfield when we heard a Mayday call requesting an urgent diversion to Gütersloh. The excited Belgian F-104 pilot stated that he had experienced a 'tree strike' and had lost part of his wing – the F-104's wing was minuscule at the best of times. We parked the aircraft by the hardened aircraft shelters and scrambled up the revetments to watch the approach and landing. Unsurprisingly it was *very* fast, the tyres smoked then burst, but the pilot somehow kept the aircraft straight and successfully engaged the barrier at the eastern end of the runway, still at some considerable speed. We drove out and picked up the pilot, later dubbed 'cool hand Luke', and took him for a restorative coffee. He explained that his formation leader on a ground-attack exercise had initiated a low fly-past of a forward air controller's position, pulling up very late leaving wingman 'Luke' even lower. His aircraft ploughed through the tops of several trees, parts of which were still embedded in the fuselage and wing. Despite losing part of an aileron and a drop tank, he had managed to recover the jet.

OFF-BASE OPERATIONS

The prevailing wisdom in the Cold War was that all NATO main bases would be targeted and highly vulnerable to early attack by the Warsaw Pact, notwithstanding the area and airfield ground-to-air defences. But the unique characteristics of the Harrier, to take off from very short, rudimentary strips and land vertically, allowed our force the option of deploying and operating off-base, from roads and rapidly prepared rural sites; a fully armed Harrier needed just 400 yards of prepared surface to get airborne in 25°C. We practised this demanding and exhilarating option at least three times each year, for two weeks at a time, often under the scrutiny of the NATO evaluation teams.

First, the site planning team would choose and book the location; army training areas, such as Bergen-Hohne, Detmold, and Paderborn were often used but farmers' fields and local roads in the flat plains of north Germany were also options. The Royal Engineers would provide the vertical landing pads and build a take-off strip out of pierced aluminium planking ('tin' to the lads) where there was no existing hard base. Then the Royal Signals team would move in and set up the necessary on-site communications and links. We became very close to these two army units and could not operate off-base without them. They worked all hours, fixing the planking that we regularly trashed and maintaining comms in a hostile environment of remote woods and occasional bad weather. They tended to be first in, last out, and were a great bunch of chaps.

Each of our two squadrons would be allocated six flying sites, with two reserves ('step-ups'), and there were force logistics parks. Each site would house around 200 military personnel. The RAF Regiment worked with the army to provide security, and the supporting teams of engineers, logisticians, air traffic controllers, ground-liaison officers and many more would convoy out prior to the aircraft flying in, nominally six per site. We landed vertically on the pad, taxied to the hides, and prepared for an intense period of flying. We flew between two and six sorties per day, with cockpit turnarounds and a break in the middle. We started slowly, assessing the specific problems presented by each individual site before raising the tempo. Taxiing on the strips of planking was tight and needed precise marshalling. 'Bog-ins' were not uncommon on the more agricultural sites and the ultra-sturdy undercarriage proved its worth; take-offs could be bumpy, and approaches to vertical landings needed to be accurate to avoid blowing away tents, hides, equipment, and the tin. For the aircrew, finding the site in the first place could be a challenge – the early on-board navigation system was rudimentary, the sites were camouflaged and built to be hard to locate. It was not unknown for a returning aircraft to make a second run or divert if too low on fuel. And the weather was always a limiting factor.

Each site was led by a flight commander and the field wing operation centre was headed by the force commander. Tasking ramped up as the exercise progressed. Responsibility for a site, the operations and the safety and welfare of the men and women was down to the site commander, 24/7 for the duration of the deployment, a heavy task for relatively junior officer ranks, even with the strong support of his teams. Engineering in the field was challenging, particularly deep maintenance – the engineers worked in whatever weather was thrown at them, night rectification in the cramped spaces was complex, and tiredness was a constant concern. However, the engineering teams were skilled and well managed, and proved their expertise daily, sometimes even changing an engine in the field overnight – no mean feat.

We lived in 12-ft square tents, six to a tent, and were fed superbly by the mobile catering support unit. Ablutions were basic, 'long drops' common, and could pose hazards in the dark which were not covered by Pilots' Notes. The army was good at loaning out their shower houses, if local. On one remote site the word got out that there was a rather select German

A Harrier GR3 of 3 Squadron taxies from its hide.

Engine change in the field.

health complex nearby, with pools, saunas and facilities. The news travelled at lightning speed and within hours the place was swarming with excited techies in camouflage kit carrying MOD-issue green towels. That's one reason why we employed a German liaison officer; she worked hard to smooth troubled waters that day.

The off-base concept required not just concealment but mobility, and each site was required to be ready to move to another location within hours. So, when on high alert, our portable canvas and steel beds were cleared away every morning before met brief, and the few possessions packed ready to bug out. NATO's multi-national tactical evaluation team used the off-base exercises to test the efficiency and sustainability of the operation, and we prided ourselves on being consistently awarded the highest grades. But they were long days. To be fair, the off-base work, though invigorating and fun, was knackering and the next week or so back at base was usually spent adjusting to a normal routine, sorting out the various snags on the aircraft, catching up on missed sleep, and washing all the muddy kit and 'green worms' (sleeping bags). Returning to base operations seemed almost luxurious.

START OF THE THAW – 9 NOVEMBER 1989

In the latter years of the 1980s we all became aware of the softening of relations between East and West, and there were indications that the Cold War that had dictated our professional lives for the previous two decades was about to end. On the night of 9 November 1989 at Gütersloh we felt the vibes that an historic event was close. That day I flew a fifty-minute sortie in a GR5 as part of a four-ship bounced simulated attack near the Weser River, a stone's throw from the IGB. The next day we woke to the news that the Berlin Wall was being dismantled, and a new era began. The massive fences, walls and no-go zones that had divided Germany for over forty years were razed to the ground. Just days later I took my family for a long-planned break in the Hartz mountains; Bad Harzburg in the 'old' West was, as ever, colourful, vibrant, and visibly wealthy; just a few kilometres over the now defunct border, the medieval town of Wernigerode was grey, dull, with near-empty shops and an aura of decay. The difference was stark. Our German hosts sensed reunification, the autobahn from Helmstadt through to the Ruhr became increasingly clogged with slow and smoky Trabant and Wartburg cars driven by curious 'Essies', the newspapers were full of talk of glasnost, and Europe oozed optimism. The excellent 250-foot low-flying system in Germany quickly came under scrutiny and ended within a year, forcing our ground-attack training back to the UK, and base closures and unit moves home became the way forward. Interestingly the quality of the Warsaw Pact training and equipment, so mightily praised in our regular morning intelligence briefings, proved to be far less impressive on close inspection; whilst the tactic of painting the opposition as highly competent may have spurred greater defence spending in the west, and motivated all potential war fighters to up their skills, over-egging the pudding made many of us very cautious of intelligence claims thereafter.

So, the halcyon years of work hard/play hard on the continent started to slip away. Not all had gone perfectly; sadly, we lost some excellent aircrew to accidents, living overseas in a bubble did create some domestic tensions and turbulence, and at the macro level the financial investment of basing in Germany was massive. The IRA were occasionally active against British service personnel serving in Germany through the 1970s and 1980s, killing at least six and injuring more by bombing and shootings, and this impacted on free movement to a limited extent. But most who experienced the two decades in uniform in Europe look back with great pride and affection on a time of strong professionalism and achievement. We worked with many hugely talented and able people and shared the same ideals. And, in the final analysis, the Cold War was a long and at times very tense confrontation that ended smoothly and peacefully.

Job done.

CHAPTER FIVE

THE LIGHTNING IN RAF GERMANY

SIR IAN MACFADYEN

IN 1960 THERE WAS A push to try and withdraw all RAF fighter aircraft from the recently formed command of RAF Germany. With the shooting down of Gary Powers in his U-2, relations with the Soviet Union soured. A new NATO requirement arose called 'Jack Pine': the need to confront, if necessary, Warsaw Pact fighters trying to block the Berlin air corridors. More on this later. At the same time the Foreign Office was also keen to maintain good relations with the Federal Government in Bonn and backed the idea that RAF fighters should remain in theatre. The outcome was that one Hunter, and two Javelin squadrons were retained although even these were planned to be withdrawn at the end of 1962. SACEUR later intervened stating that gun-armed-only fighters were unacceptable and in future missile-armed fighters were essential. Plans were therefore put in place to deploy two Lightning squadrons to RAF Geilenkirchen, then a Javelin base. This was later revised to initially base one squadron at RAF Gütersloh with a second at Geilenkirchen as there was insufficient ramp space at the time at Gütersloh for two.

Ian Macfadyen boards his Lightning in Cyprus.

In 1965 I was aged twenty-two and enjoying my first flying tour on 19 Squadron at RAF Leconfield flying the Lightning F.2. I was on detachment with a small force of four Lightnings in Cyprus that summer on QRA duties because of the ongoing dispute between Greece and Turkey. The resident Javelin squadron in Cyprus was away on deployment to Rhodesia. I learnt while I was there that 19 Squadron was to be the first to deploy the Lightning to RAF Germany. On my return to the UK, it was time to get ready for the move.

The day dawned on 23 September 1965 with twelve Lightnings lined up and ready to fly to Gütersloh. I will not forget our arrival there. It was a big event in the history of RAF Germany operations to bring supersonic aircraft into the country for the first time and there was much press interest. All aircraft departed Leconfield successfully, and about forty-five minutes later we were in three groups of four, lined up in echelon starboard and running in for a left-hand break. This is when things started to go wrong. My CO, Wing Commander Brian Cox, had failed to check the circuit direction in his planning. Circuits at Gütersloh were always right-hand. With twelve aircraft approaching the airfield, and getting short of fuel, there was no time to change the formation to echelon port. We all thus did a left-hand circuit which did not please our new station commander, Group Captain David Evans. We had arrived! Our sister squadron at Leconfield, 92, was to follow on to RAF Geilenkirchen in December 1965, to remain there until January 1968 when 18

Lightning F.2s of 19 Squadron.

Squadron, flying the Wessex helicopter, was withdrawn from Gütersloh leaving space for the second Lightning squadron.

Our squadron accommodation at Gütersloh was in a former Luftwaffe hangar which had barely changed since the end of World War Two when it had been used by Ju-88 night-fighters. Internal hangar space was small. Gütersloh was already a busy base with 2 and 4 Squadrons flying Hunter FR.10s, as well as the Wessex Squadron. The station was about to get a whole lot busier.

Our principal peacetime role was to protect the air sovereignty of the northern half of West Germany. To do this we held two aircraft on high alert fully armed by day and by night. This was known as Battle Flight. As an already operational air defence squadron, we only required a short time to work up before starting to share Battle Flight duties with the one remaining Javelin squadron based further west at RAF Geilenkirchen. This continued until January 1966 by which time 92 Squadron had arrived at Geilenkirchen. Lightning Battle Flight was held at Gütersloh with, when required, a detachment of aircraft from 92. Initially, we held Battle Flight from the front of our hangar, but two purpose-built new small hangars were constructed by mid-1966 near to the entrance to the western runway to provide all-weather accommodation for the two fully armed aircraft, pilots and ground crew at ten-minute readiness, known as RS10. A system called telebrief enabled the pilot to get direct instructions from the senior fighter controller in the radar station before and during engine start. Apart from an occasional launch to test the system, Battle Flight aircraft were 'scrambled' relatively infrequently and usually because a light aircraft had strayed into the ADIZ, a safety border area about 30-nm wide along the Inner German Border. Identifying such as a Cessna 152, flying at about ninety knots, was difficult when the Lightning's minimum practical speed was double that.

Although the first few months of settling in included little night flying, day-to-day flying training soon morphed into a flying programme starting around 08:00 and working through until most aircraft were unserviceable. The aircrew were divided into A and B Flights swapping around each week, with the late flying programme starting with a weather brief at around 16:00. To support this, the squadron ground crew were divided into day and night shifts, swapping every two weeks.

The flying programme consisted mostly of pairs of aircraft on mutual medium- and high-level interceptions with one aircraft initially acting as a target and then changing to attack. Aircraft would split by about forty miles and then turn inbound towards each other for either about a 180- or ninety-degree attack. The 'attack' pilot would use his AI.23 radar to find the target, work out its height and heading and then conduct a stern intercept to a position where one could simulate the firing of the Firestreak air-to-air missile. General ground control and monitoring would be by a Luftwaffe-manned radar control

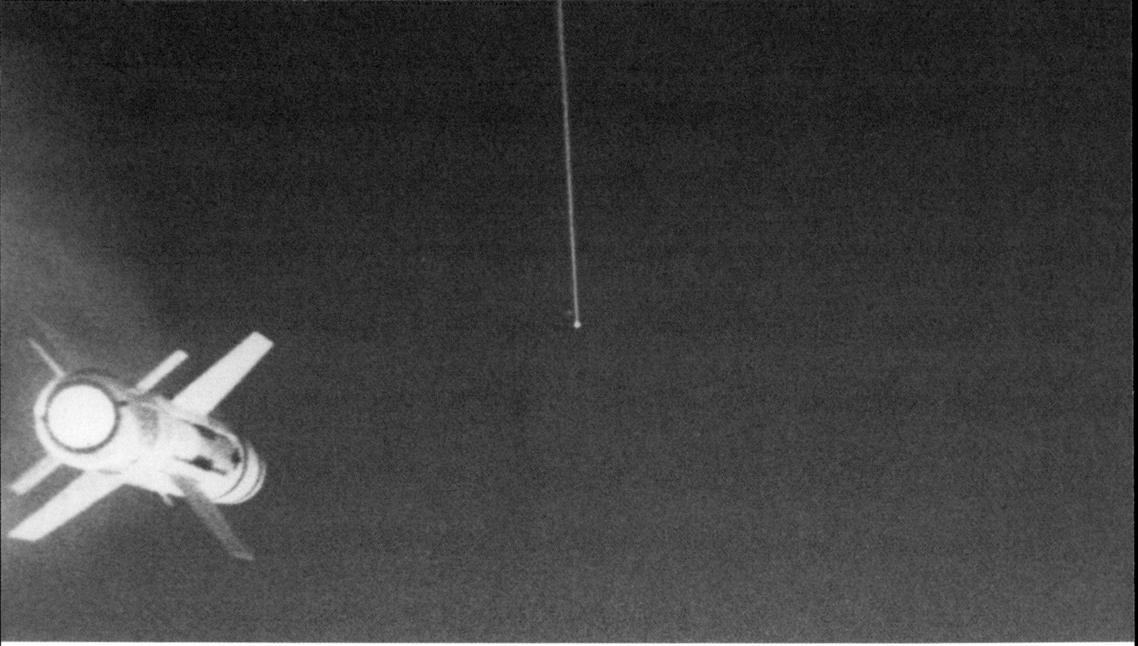

A Firestreak missile on its way to the target.

and reporting centre, mostly at Auenhausen, callsign 'Backwash'. Sometimes the fighter controller would control the intercept and at other times, especially on a large exercise when multiple targets were to be seen, the fighter controller would broadcast details of an incoming raid, for fighters to find their own targets. An important part of the work was simulating trying to identify a target by night using closure on the aircraft's AI.23 radar down to the minimum of 200 yards. This was practised in daylight initially for safety reasons but once a pilot had been judged competent, he would move on to the same by night. It was a near impossible way of identifying an unknown target, especially if the target was unlit, unless there was good moonlight, since such things as night-vision goggles were a thing of the future. Intercepts were always filmed so that the pilot could be debriefed on the results and improve his skills. If one aircraft went unserviceable on start, the other would often go off on 'freelance PIs' (practice intercepts) to find targets of opportunity. The Luftwaffe F-86 Sabre Wing at Ahlhorn was a source of many possible targets and the ensuing dog fights were always exciting, if not actually authorised! Another testing form of all-weather training was flying against a Canberra radar frequency jammer on Exercise Corncob.

With a growing Soviet high-altitude reconnaissance and bomber threat, practice intercepts on targets at 50,000 feet and above were part of the training programme. For this, one Lightning would fly at 50,000 feet at speeds up to Mach 1.1, with the interceptor Lightning going supersonic to Mach 1.7 at 36,000 feet and then zoom climbing up to the target. With little fuel available, it called for considerable intercept precision, with help from our Luftwaffe fighter controllers. Supersonic flying overland was acceptable in those days. In May 1966, an inter-ATAF air intercept competition was held at RAF Geilenkirchen to try and improve standards. One morning a spare Lightning was required urgently at Geilenkirchen and Flying Officer Duckett logged a fifteen-minute supersonic flight for the 100-nm journey! He would have had little remaining fuel when he arrived.

To test the full operational system, and to give experience to aircrew, the squadron deployed from time to time to the missile practice camp at RAF Valley in Anglesey to fire missiles on the Aberporth Range. The first such deployment for 19 Squadron was in October and early November 1966. I was fortunate to fire a live warhead (most were not, to ensure the Jindivik unmanned drone towing a flare was undamaged). I was in the two-seat Lightning with Flying Officer John Rigby. The resulting film showed a spectacular success with the towed-target flare destroyed by a direct hit. In the commentary I was making we both initially failed to recognise that a primary alarm had gone off as we fired the missile. Fortunately, the caption went out without any further difficulty.

Such were the complex workings of the Lightning that things quite frequently went wrong. The radar was unreliable, and such mishaps as hydraulic or re-heat failures were not uncommon. The relatively short runway at Gütersloh provided other hazards. It was less than 7,400 feet long and in the event of a brake parachute failure, there was always

Pilots of 19 Squadron at RAF Valley for a missile practice camp, 1966. Ian Macfadyen is on the far left.

a risk of engaging the runway overrun barrier. The normal procedure was to jettison the failed parachute and go around for another landing. 'Precautionary' landings were announced on the radio if the pilot wished to remain on the ground in the event of a parachute failure. This alerted the crash crew to a possible barrier engagement. Strong crosswinds could tear all the rubber off a brand-new tyre, the pressure of which was set at 330–350 psi, in just one landing.

Many of the more junior squadron pilots were also able to fly one of two two-seater Hunters that were allotted to the station flight and used for all kinds of liaison work. It helped with building up flying experience. These aircraft were sadly withdrawn from use in August 1966.

RAF Gütersloh was the nearest RAF Germany airbase to the IGB thus shortening the time taken to reach any incursion by Warsaw Pact forces. In other ways it was not an ideal location, not only as a more vulnerable air base than most but also because it was in the middle of the extensive surface-to-air missile (SAM) belts which ran north/south down West Germany. The forward low-level belt consisted mostly of Hawk missiles with the Nike Hercules providing the rear high-level threat belt. These were manned by German, Dutch and Belgian crews. In wartime, this belt would complicate procedures to prevent a blue-on-blue engagement. On exercise (and in war), we thus initially flew west to get beyond the SAM belts and set up a fighter CAP rather nearer the 'clutch' airfields as they were known: Laarbruch, Brüggen, Wildenrath and Geilenkirchen.

At the end of World War Two, it was agreed that Berlin should become a quadripartite city and so it was divided into American, British, French, and Soviet sectors, with three air corridors from West Germany agreed and set up across East Germany to and from the city. It was these corridors that had enabled the Berlin Airlift to save West Berlin from starvation in 1948. Subsequently, due to a further rise in tension in 1960, the three Western allies agreed that, in the event of any future interference by the Soviets, the West would be prepared to protect the corridors from any unwanted intruder. This required a joint operation, known as Live Oak, with each nation providing fighter aircraft. To best coordinate the operation the former RAF base at Fassberg, close to the IGB, was chosen as the mounting base. It was near to the shortest middle corridor to Berlin but could also protect the other two corridors to the north and south. The exercise was carried out about every two years when detachments of Allied fighters would gather to work together in providing escort operations to transport and other aircraft using the corridors. The idea was that up to four fighter aircraft from each nation would provide escort cover working as a collective group. On exercise, it was all flown away from the Berlin air corridors, but in sight of Soviet radar stations, with up to twelve Allied fighters, American, British, and French, who would form up as an escort group. This was to demonstrate that any air threat in the corridors would be countered, although as they were only 18-nm wide it would have been a cumbersome operation.

A NATO exercise, called Taceval, had been set up in the early 1960s to test the operational readiness of all NATO air forces. In the early days it was a one- or two-day exercise, later developed into something far more sophisticated. No. 19 Squadron was tested at RAF Gütersloh for the first time in May 1966, and then again in April 1967 at the end of which the squadron flew a diamond nine formation as a mark of its robustness after an intensive day of flying. The third such evaluation was in February 1968 shortly after 92 Squadron had moved to Gütersloh. The RAF generally did well in these tests.

Exchanges between squadrons were a way of fostering goodwill and the sharing of ideas on such things as tactics. No. 19 Squadron had a two-week exchange with the USAF 525th Squadron at Bitburg in April 1967. This squadron, equipped with the Phantom F-4D, had a distinguished history, and most of its aircrew had considerable Vietnam experience. That was sobering as we had no combat experience. However, much as we admired and, in its way, envied their operational experience, we found the USAF control of air traffic to be extremely rigid. On one occasion, on a large exercise, two Lightnings were held at the take-off point for so long that they were forced eventually to taxi back to refuel.

Besides exchanges, there were many open days amongst Allied air bases which gave opportunities to show the flag. Flight Lieutenant Bill Wratten was the RAF Germany Lightning solo demonstration pilot for 1966 and 1967. Other pilots either escorted him with a spare aircraft for a display away from Gütersloh or acted as a static display machine. A visit to West Germany by the commander of Soviet Forces in East Germany took place in February 1966. He landed at Gütersloh and was met by a British Army Guard of Honour. The escorts were each well over six feet five inches…and he was little more than five feet! After the parade our CO, Wing Commander Brian Cox, led a four-ship display. It began with a well-known reheat take-off, snatching over 3 g to the near vertical as one reached 250 knots. It was an impressive start, demonstrating the Lightning's remarkable performance and agility. Gütersloh itself had a public open day on 27 May 1967. No. 19 Squadron again provided a four ship formation display. The leader was again Wing Commander Cox, with Flight Lieutenants Al Blackley, Barry Grainger, and me as Nos. 2, 3 and 4.

The medium- and high-level intercept roles ignored the growing low-level attack threat. For the Lightning, the problem was that the AI.23 radar did not work well in ground clutter. Indeed, the nearer to the ground we were, the more was the difficulty in seeing a target on radar at all. Nevertheless, the new threat forced us to begin to fly more at low level. The best tactic, in early 1968, seemed to be to do a north/south orbit looking visually towards the threat direction. Since we were flying underneath the high-level Hercules SAM belt this worked adequately. The other problem with the Lightning, its short range, meant

No. 19 Squadron's hangar.

that loitering at a low altitude used fuel at a much faster rate. We thus developed a tactic of shutting down one engine to preserve fuel, only starting it when the threat so demanded. The entry into service of the Lightning F.2A was to help change this dramatically.

The F.2A, often described as the best Lightning of them all for its ability to have two or four internal 30-mm guns (unlike the F.6 which had only two in the fuel pack under the fuselage), was a much-modified F.2, rather than a new-build. A programme began in 1967 to send the F.2s back to BAC at Warton to be so modified. The improvements were a much enlarged and fixed under-fuselage fuel tank, an extended fin, and a new cranked wing with slightly increased area. The modification programme took about five months for each aircraft. The first F.2A arrived on 19 Squadron in early 1968 just before I was posted back to the UK. I therefore only flew this greatly improved aircraft a few times. Sortie lengths were typically at least seventy-five minutes as opposed to around forty-five in the F.2 thus greatly improving the 'on task' time.

The Lightning has been regarded as one of the more remarkable RAF aircraft with a tremendous performance and great handling qualities. It was my pleasure and delight to fly over 800 hours in this fine aeroplane, mostly in RAF Germany, before moving on to pastures new.

CHAPTER SIX

LIFE AT RAF BRÜGGEN IN THE COLD WAR

ROB WRIGHT

It is autumn 1971 and I have just been declared combat ready on 17(F) Squadron based at RAF Brüggen. I am twenty-four years old and for the first time I am sitting on fifteen-minute readiness in Brüggen's nuclear QRA building. Outside in revetments sit three Phantom FGR2 strike-attack aircraft, each armed with an American tactical nuclear bomb closely guarded by armed USAF personnel.

With the crews from the two other Phantom squadrons based at Brüggen, my navigator and I are carrying out a duty that mirrors the many years served by our Canberra forebears. The difference is not that the individual spirit and commitment of our crews has changed, but that now we are equipped with a truly multi-role high-performance aircraft, the newest in the RAF inventory and one that requires aircrew expertise across a hugely expanded range of tasks and capabilities.

Rob Wright with his USN Phantom.

This is an aircraft that has a Mach 2 capability, can perform in all attack roles as well as that of the long-standing strike (nuclear) role and has set every world record for speed, time to height and maximum altitude. It is fully equipped with state-of-the-art equipment including look-down, shoot-down pulse Doppler radar linked to long-range Sparrow air-to-air missiles and short-range Sidewinder infrared-heat-seeking missiles. We can navigate with a self-contained and unjammable Ferranti inertial navigational/attack system and are equipped with electronic jamming systems and various self-defence measures.

On the ground, it looks like a triumph of thrust over drag, but this very much belies the aircraft's capabilities. Whilst the Mach 2 capacity is of no relevance on a heavily loaded aircraft at low level, the thrust (40,000 lbs total) from two Rolls-Royce Spey engines readily enables speeds of over 550 kts at low level. It does not have the longer range of the Canberra, but it can now fight its way to the target, defend itself and return home. It can be armed with SNEB rocket pods, 1,000-lb bombs, an immensely powerful Gatling gun, the SUU-23, capable of firing 6,000 rounds a minute, and with a forward-looking F95 reconnaissance camera and air-to-air missiles, this is indeed a formidable aeroplane.

We are fortunate to have this very capable and combat-proven aeroplane in our inventory following the political debacle of the cancellation of the TSR-2 strike aircraft,

A Phantom FGR2 of 17 Squadron over Germany.

the cancellation of the P.1154 project to replace the UK ground-attack capability, and the planned procurement of the F-111 bomber. After this litany of failure, it was decided the RAF would take over the Fleet Air Arm's Buccaneer and Phantom FG1s and buy additional Phantoms F-4Ks from the United States. These would be modified with British equipment, in particular the Rolls-Royce Spey bypass engines, giving a total UK content of forty-five per cent of the aircraft. A purchase of new-build Buccaneers for the RAF was also approved.

I had converted to the Phantom at RAF Coningsby early in 1971 after a time spent in the Middle East flying the Hunter in the day-fighter ground-attack role. Whilst the Hunter was a superb aeroplane to fly, it was a revelation to become acquainted with the Phantom and to become proficient in all its capabilities, air-ground of course, but also air-to-air combat and intercept training when weather precluded operating at low level; this allied to an introduction to a very select breed, called navigators, who sit in the rear seat and control all the intricacies of the navigational and weapons systems.

The newly formed Phantom squadrons were crewed in the main by ex-Lightning and Hunter pilots and by ex-Canberra and Vulcan navigators, interspersed with a considerable number of first tourists straight out of training. We all had a great deal to learn, and I recall long, but friendly, discussions on how, as individual crews, we would develop the all-important crew cooperation which was the key to success in the Phantom. A perennial debate in the early months of these developing relationships was the carrying of maps. Having had a single-seat tour I had always carried one, but my navigator told me that as he was doing the navigation there was no requirement for me to carry a map at all. This was to bite us in our crew time together, more of which later. Eventually, we agreed an affable compromise on this in that he as the navigator would have his, but I always carried a slightly less detailed one to keep abreast of situational awareness, particularly when leading formations of up to eight aircraft.

Over time, I developed a great and increasing respect for these talented aircrew and was of the firm view that if we had to go to war, particularly against the backdrop of the very complex NATO scenarios in Germany in the 1970s, I would want to be in a Phantom with a navigator in the back.

Operating in western Germany in the 1970s was predominantly a low-level environment, 250 ft in most of the low-level airspace and lower on the tactical ranges where we carried out a variety of attack profiles: high-speed level bombing, toss bombing, and all the conventional dive profiles with bombs, guns, and rockets. Winter weather dictated that we often flew in relatively poor conditions, as once stuck above cloud it was problematic getting back down to low level. This is where crew cooperation became so important, and realistic training the key to success. All of this was undertaken within the complexity of NATO procedures, the numerous station and NATO-wide exercises that were programmed throughout the year, in the panoply of Minivals and Tacevals (major evaluation tests – fail and you are finished) and squadron exchanges.

In a week on exchange with a Canadian F-104 Starfighter squadron based at Baden-Söllingen in southern Germany, I had the opportunity to fly in the rocket-ship F-104 (in the front seat which I discovered was against their regulations, but we got away with it). When this squadron exchange was returned at Brüggen, I did reciprocate with a ride in my Phantom, but with "apologies back seat only".

We all took great pride in our personal weapons' scores which led to much competition, on the squadron, inter-squadron on the station, internationally on squadron exchanges and in NATO competitions, such as the prestigious Weapons Meet. It was in this international competition in 1972 that a close friend on 17 Squadron won top-scoring pilot and the Americans could not understand how a relatively young junior pilot with much less than 500 hours on type could get such high scores.

Add to all of this, detachments to Anglesey in North Wales on missile practice camps to fire Sparrow and Sidewinder missiles, and occasional weekend trips to the Mediterranean (Lone Rangers) to prove that we can deploy anywhere in southern Europe whilst enjoying the experience at HMG expense. All of this meant that life was as busy and as interesting as it can get.

A major contribution to our combat effectiveness was the training and close relationships we developed with the Harriers at Wildenrath and the Buccaneer force and reconnaissance Phantom Squadron at Laarbruch, our sister 'clutch' stations a few kilometres away. This training entailed not only regular 'bouncing' of each other's formations (usually authorised and excellent training) but was epitomised by competition in the renowned Salmond Trophy bombing and navigation exercise competed for every year; and not to forget the annual Aircrew Soccer Competition – all events equally hard fought between all wings and squadrons.

A regular and popular deployment was to Decimomannu in Sardina for intensive weapons training on Capo Frasca range where we could practise all our weapon deliveries whilst engaging in competitive banter as to who was 'Top Gun'. It was here that my navigator and I fell foul of our agreement over map carrying. We were leading a four-ship with the squadron boss as No. 2 and, unusually, we had full wing tanks. To ensure we were within 'g' limits for dive bombing, we had to burn off fuel on a thirty-minute low-level route prior to entering the range routine. At Point X-ray (prior to the holding point to entering the range) we planned to set off on our low-level route avoiding towns on our carefully planned navigational exercise. As we approached X-ray, I asked for the heading and a very red-faced navigator (we had rear-view mirrors in the front!) confessed he had

No. 17 Squadron at Decimomannu. There is a 14 Squadron interloper in the aircraft line up.

forgotten his map. A very smug pilot said: "No probs, I will use my map." (I could be magnanimous in the extreme…) Horror! I realised that I had also forgotten mine. My navigator wanted to hand over the lead to the boss, but I replied, "no way". I called the formation into a tight finger-four formation and flew the route from memory avoiding all towns.

The boss was an ex-Hunter squadron commander who I knew from Bahrain and in the debrief he was delighted to have engaged in some 'wadi flying'. "Excellent idea Rob."

In that debrief I did not reveal our mutual faux pas but instead concentrated on the weapon scores – our *raison d'être.* Thankfully, as the boss had a bad series of scores, he was a little distracted during the debriefing. My navigator was most grateful, and crew cooperation (and his reputation) was saved.

Back at Brüggen, life continued much as any UK town with a population of over 10,000. Operationally it was always at the top of its game. Like all 'clutch' bases, along with Gütersloh further east, we were very conscious that we were in the front line of the Cold War against our enemy the Warsaw Pact. The base was extremely well camouflaged, surrounded by barbed wire and heavily guarded, protected by a squadron of Bloodhound surface-to-air missiles, all aircraft were tucked away in revetments on exercises but operated from the flight line on day-to-day operations. And all this a ten-minute drive from our mess and our homes. A fine competitive spirit with many enduring friendships, still in being today, a mutual respect epitomised by squadrons putting on a barrel of beer on Friday nights for the other rival squadrons. Problems arose on one occasion when 14 Squadron put the beer barrel in the tree outside the officers' mess, but it was all consumed within the hour.

For families there was easy access to the medical centre and there was a well-equipped hospital at nearby Wegberg. Brüggen had shops, cinemas, excellent sports facilities, a maintenance unit, a railway line for bringing weapons in from the nearby bomb dump and a superb eighteen-hole golf course (ostensibly built to disperse our weapons on small, manicured areas called greens with a pin and flag marking the centre!).

Brüggen's leadership was of the highest calibre. All three station commanders that I served under were very approachable and flew regularly with the squadrons. Ordinarily, aircrew were wary of contact with the top brass (could spell trouble?). I flew Group Captain Curtiss (a wartime Bomber Command and Javelin navigator) in the back seat of the Phantom.

We were all entertained (flying officers to drinks, squadron leaders to a buffet supper, wing commanders to dinner) by Group Captain Harcourt-Smith where he ordered three couples from 17 Squadron back into the lounge of Brüggen House; we had attended his drinks party and were at the back of the queue to leave (reluctantly) at the allotted time of 20:00 but he wanted to continue. We left at 22:00. I also led missions with Group Captain Peter Harding on my wing, a former helicopter squadron commander, having been tasked with introducing him to the low-level attack role of his Phantom Wing. Flying with the squadrons was an effective leveller and helped mutual understanding between the commanders and their young charges.

In our overseas environment we were largely immune to the many dramas back in the UK which was facing a time of poor salaries, strikes and a four-day working week. The land of duty-free, tax-free cars, skiing in Austria and trips on the continent cushioned us from those hardships back home and we were relatively well off compared to many of our UK-based colleagues. The complaint from them was that you could not be posted to Germany unless you had served there previously!

This was a Cold War that everyone at Brüggen took seriously. Ironically it was a very stable period punctuated by periods of tension as the hooter blew in the early hours, and while we were never really tested in conflict, the scenario was as real as one could imagine. All of this provided an excellent and effective underpinning of the mantra 'Work Hard Play Hard'.

Inevitably in this hard fought and dynamic flying environment there were some casualties. Sometime after I retired, I visited the Armed Forces Memorial in Staffordshire which commemorates those who have lost their lives in service since World War Two. At this impressive memorial I looked up the inscribed names of close colleagues that had lost their lives flying in Germany. In the 1972 column, in sequence, I read the names of five aircrew based at Brüggen, three of whom were on my squadron and all of whom were killed, flying at low level, during a tragic and very sad three-month period. Looking at the adjacent column depicting 1973, I noticed the names of two Harrier pilots from Wildenrath, both good friends from training and my Hunter time and who had also lost their lives flying in Germany. Inevitably there were numerous close calls in this dynamic environment: I watched a Phantom take off at Brüggen and saw it climb very steeply as the wing-fold mechanism (a throwback on some aeroplanes from the original navy design) had not been locked, with the result that the outer wings folded on rotation and I looked on as the crew ejected and landed by parachute back onto the runway. Not my squadron by the way.

In 1973 I planned a trip with my family to visit West Berlin by car. This proved to be quite a daunting and challenging experience not only in the execution but also in the planning. All travellers needed to have proper documentation and all car and insurance documents had to be in perfect order. Because of the role I held as a RAF squadron pilot this was particularly important. We knew that every document I submitted would be subject to intense scrutiny and that any information would be copied and filed on the personal files that we understood were held somewhere in the depths of Warsaw Pact Records Office. This proved to be the case at Helmstedt checkpoint where I handed all my documents through a wire shutter, no one visible just a hand, and from outside sitting in the car my wife could see a group of officials frantically photocopying all the documents

A 17 Squadron Phantom makes the obligatory visit to the Möhne Dam.

I had just handed over. A camera mounted above the television in the foyer filmed the process – with me as the only star. The tension continued as we travelled east down the autobahn to West Berlin where we were tailed by a Volkspolizei (VoPo) police car. The VoPo was the East German police force consisting of over 250,000 personnel to keep control of the civilian population. They tracked me for several kilometres from very close line astern (couple of metres) before peeling off onto a slip road, and then reappearing at the next turn off where they were pointing a radar gun to catch you speeding – which was one's natural reaction to all this intimidation once the VoPo had broken away. If you arrived at the West Berlin checkpoint before a certain time lapse you would then be fined.

Travelling down the autobahn and looking to the north, close to Magdeburg, I spotted the surface-to-air missile site, which was one of our pre-planned targets (known as Option A) which we would pre-emptively strike in the first hours of any conflict. Probably taking target study a bit too far methinks.

That aside, this trip was a memorable experience and one which brought home the stark symbol of the division between east and west during the Cold War. Staying in the very comfortable Edinburgh (Military) Hotel in West Berlin, at ridiculously low cost, illustrated the desire to show the east how well off and affluent we were compared to the surrounding population, and that we were not to be intimidated. That did not stop the Soviet guards at checkpoints adopting a strict and intimidating demeanour, however, their main aim was to cadge cheap western cigarettes. "Nyet" was the only Russian word I used and which they understood and accepted, albeit reluctantly. Overall, travelling by car was not a simple or straightforward endeavour. It required meticulous planning, patience, and a keen awareness of the political and security situation. It was an experience I will not forget and brought home the reality of the Cold War and the part we played in our defence.

On reflection, my three years as a Cold War pilot overseas provided a memorable and instructive foundation for the rest of my career. The close friendships established in those halcyon days remain with us all – we had 'the best of times' is our battle cry in retirement. Underneath the Brüggen crest are the words 'To Seek and Strike', a motto that has accompanied the station since it was built in 1953, operating different aircraft with different people, but always with the same commitment and pride, the same determination and professionalism and above all with the same spirit.

Brüggen has played a major part of my life: I served there for a total of ten years as the pilot officer camp commandant of CCF and ATC Star Camps, as a junior pilot and adjutant on No. 17(F), as a squadron commander of IX(B) Squadron Tornado, and as the station commander. I met my wife at Brüggen (a WRAF officer) and my son was born at nearby Wegberg Hospital. These times are etched in my memory, and I reflect that, two

Rob Wright 'traps' on USS *Saratoga.*

years after leaving 17(F) Squadron, and via a short tour at RAF Brawdy on the Tactical Weapons Unit as a weapons instructor, I was once again in a Phantom, but making my first deck landing on USS *Saratoga* in a US Navy F-4J. I was the fighter weapons instructor on USN VF-101 Squadron based at Oceana, Virginia Beach, still overseas and the same Cold War, but with a different perspective and different potential scenarios in the Atlantic and the Mediterranean. But that is another story.

CHAPTER SEVEN

SUPPORT HELICOPTERS IN GERMANY AND BEYOND

TONY STABLES

I joined 18 Squadron operating the Wessex helicopter and based at RAF Acklington in April 1969 having completed my first tour on 78 Squadron, also operating the Wessex, at RAF Sharjah in the Trucial Oman States. No. 18 Squadron had moved from RAF Gütersloh in Germany to RAF Acklington as part of the 'Healy' defence cuts in the 1960s. We retained a two-aircraft standby role to respond to any incursion of the IGB but given that our icing limits at that time prevented cloud penetration at temperatures below plus ten, our transit capability was restricted to day VFR, although our flying limits were a generous 100 feet cloud base and half nautical mile visibility. Only fog would stop us.

Tony Stables with an 18 Squadron Chinook.

I remember flying the second aircraft, the lead being a sergeant pilot (yes, we still had NCO pilots) having been scrambled from RAF Acklington to respond to an incident on the IGB but only making Newcastle Woolsington airfield on the first day.

A permanent four-aircraft detachment was established at RAF Gütersloh utilising a Rubb hangar positioned close to the Lightning QRA shed. My first detachment was in July 1969, and we mounted field deployments in support of army exercises.

Living and operating in the field was, at that time, a relatively gracious affair not least because we took with us our own cooks and stewards, but we were developing an 'off-base tactical posture' and things would quickly become more austere. Our kit was adequate, less the standard-issue flying boots which were not waterproof and not well suited to operating in wet grass and woodland.

During a subsequent field deployment in September 1969, I was captured ('volunteered' by my commanding officer) and subjected to a field interrogation by trained interrogators. The experience lasted for a couple of days and in that time included the continuous wearing of a black bag over the head secured at the neck (subsequently banned). It was very disorientating and a welcome relief when removed and my hands untied at the end of the event.

We regularly mounted flying patrols of the IGB, which was divided into four sectors in the British Zone. The lead aircraft always had a member of the British Frontier Service in the left-hand seat. They had an intimate knowledge of the border which was often difficult to identify and comprised coloured concrete posts often some distance from the more well-known fence and watch towers. We weaved in and out of fields often shadowed by a Soviet helicopter on the other side of the fence. The game was to accelerate and then flare and turn, generally watching the Soviet flash past.

My eighth sortie on the Wessex OCU in December 1967, after a total of five hours on type, had been solo engine-off landings. The idea was to set up at 1,000 ft over the piano keys on the runway, stop cock both engines and carry out an autorotation and engine off landing on the runway. Some might question the wisdom of this, given that the Wessex had two engines, but my logbook records that we were still undertaking the exercise every four months in 1969, albeit then only dual with a qualified helicopter instructor (QHI).

Our original four-aircraft detachment to RAF Gütersloh was regularly augmented with additional aircraft and crews and in July 1969 the squadron moved from RAF Acklington to RAF Odiham in Hampshire thereby reducing the transit time to Germany.

In the latter half of 1969 I was invited to call on the Air Officer Commanding 38 Group. Questioning my commanding officer as to the purpose, he opined either a court martial or the job of ADC. He favoured the former but notwithstanding in February 1970 I took up the position of ADC to AOC 38 Group.

I returned to Wessex flying in August of 1972 arriving back on 18 Squadron in September 1972, now relocated, yet again, back to RAF Gütersloh in Germany. I was to remain there until August 1977.

Crewing for the Wessex normally comprised a single pilot in the cockpit and a single crewman who generally occupied the cabin. We had little or no navigation aids and much of our flying was conducted at low level, usually at fifty feet above ground but often lower, using a 1:50,000 map and stopwatch. Given that fast jets often flew at 500/250 ft above ground it was much safer for us at fifty feet. We flew almost exclusively in the 1 (British) Corps area where the terrain and landmarks became very familiar. We aimed to make initial points within thirty seconds, and, in truth, we were very good at it. With weather limits of 100-foot cloud base and half-mile visibility, only thick fog prevented us from flying. Overhead wires were a problem and wire strikes, generally low wires taken out by the tailwheel, were not uncommon. High-tension wires, where the tops of the pylon were often shrouded in cloud did present a particular problem with some pilots adopting the very unsafe procedure of coming to the hover alongside the pylon, climbing vertically into cloud, moving sideways in cloud, and then descending. Submissions were made to headquarters to allow us to hover taxi safely under the wires, normally at a point close to the pylon providing the maximum distance between the ground and the overhead wire. Approval was ultimately given but it took a long time and not before our squadron commander had taken down some wires.

Our icing limits had been reduced to plus five degrees and ultimately plus three. The heated nose door was curve shaped with air drawn in from above, circulating through the curve until arriving at the engine intake. We had used the same nose doors in Sharjah, albeit without the nose-door heating and incorporating sand filters at the base of the door. In the early 1970s, we were in receipt of a modified nose door – the cold-running nose door – where the intake faced downwards. It was a very effective 'hoover' and was quickly withdrawn. We retained a plus three limit for flying in cloud which, given that most of our flying was at low level in northern Germany, posed no real penalty.

Flying was good and varied and averaged around twenty-five hours/month. Night flying was slightly more problematic without any navigation aids other than Decca Mk 18 which was of very limited use with a single pilot. We undertook the potentially very dangerous technique of night stream formations of up to twelve aircraft, generally without lights. It was terrifying especially if the landing aid at the destination was positioned downwind. We utilised Bardic lights which comprised five battery operated tube lights about a foot in length each resting on top of a rectangular box about nine inches tall. They were laid in the shape of a T with the boxes placed equidistant. The idea was to make an approach to the foot of the T maintaining a 'site picture'. The lights were often shrouded and thus only allowed a single direction of approach hopefully, but not always, into wind. An unexpected downwind position often resulted in aircraft required to overshoot thereby causing chaos to a timed thirty-second interval stream. Similarly, locating the lights in standing crops, especially corn, provided an added challenge. Successful night streams were a rarity.

When deployed into the field, the squadron comprised a headquarters and three four-aircraft flights. The headquarters was responsible for aircraft tasking, logistic support, intelligence, and overall coordination of effort. We were fully mobile, very tactical, and largely self-sufficient with each deployed flight having its own air and ground crews, catering team, and RAF Regiment personnel. We deployed to field locations, generally wooded where we were able to hide. The central hub of each flight was the operations truck, a modified long-wheelbase Land Rover supported by an under-command Royal Signals communications vehicle, providing secure VHF ground-to-ground communication with our headquarters. Ground-to-air communications were insecure and seldom used. The domestic equipment was lorry borne (four tonner) and overall, each flight had six to seven vehicles in total, all of which could be hidden and camouflaged. Our ground defence was provided by members of the RAF Regiment augmented as necessary by the other flight personnel; pilots included. We practised the tactical ground movement in and out of deployed field locations by day and night (without lights) including anti-ambush drills and, wherever possible, concealed vehicles in a herring-bone formation which facilitated a rapid departure – crash-out. Camouflaging aircraft in the field proved more challenging. If the ground allowed, the tail of the Wessex could be folded, two blades folded down the aircraft spine and the remaining two blades positioned fore and aft. The aircraft could then be towed or manhandled under cover. The procedure required a coordinated team effort but with practice was entirely possible. A cloth camouflage set was produced, the secret being to try and hide the cruciform shape of the blades from aerial view by mounting a tripod on the rotor hub to lift the camouflage clear of the blade surfaces. There was a lot of string.

A Wessex of 18 Squadron on field exercises in Germany.

We regularly spent periods of two to three weeks living and operating in the field often in NBC (nuclear, biological, chemical) conditions, continuously wearing NBC equipment for ten or so days and nights. We developed a concept of clean and dirty sites such that contaminated aircraft and personnel did not mix with the clean.

With no distractions, we were able to concentrate on rehearsing our wartime role and we became very good at what we were doing constrained only by our equipment – yes, issued flying boots were still not waterproof, and it was not uncommon to see crews operating in wellington boots. We were able to operate from field locations and fly at or below fifty feet over large areas, benefitting from Army Form 443, which was the instrument used to identify and secure large areas of land for training with few caveats. We enjoyed a permanent 443 area, south of Gütersloh, known as the 'playpen' which provided excellent tactical low flying, formation flying and night flying.

RAF Gütersloh in the 1970s was a very lively place, initially housing two Lightning squadrons (later replaced by two Harrier squadrons), a RAF Regiment air defence squadron and the Wessex squadron. Taceval of the station was a no-notice annual event conducted by an external team assessing all operational aspects of the station. It generally commenced with a recall of all personnel, normally timed at 04:30 to 05:00 to generate the station to a war footing. Given that our role was 'off-base', we responded to the call-out and thereafter deployed by air and road to exercise war sites. Looking back, we seemed to be ever either on station exercise (Minival) or deployed on exercise in support of 1 (BR) Corps. Notwithstanding, life was good for us living in the 1970s in Germany compared to the UK (three-day week, fuel shortages etc.) and we benefitted from tax-free alcohol, tobacco (far more people smoked then), cars and local overseas allowance. Unsurprisingly, we enjoyed a thriving social life and the opportunity to travel extensively throughout Western Europe. Travel to Berlin was a unique experience by road or rail through East Germany with freedom of pedestrian access into East Berlin, in uniform, a rather strange feeling. We worked hard and played hard and life was really good.

We benefitted from the challenging opportunity to practise mountain flying techniques in the Bavarian Alps positioning at the German military bases of Landsberg or Kaufbeuren. It was on one occasion in the officers' mess in Landsberg that we were introduced to 'hubshrauber shnapps' or helicopter schnapps. Made by soaking chilli peppers in schnapps for around three months, it produced an amber-coloured liquid which when drunk produced a fiery hot, tingling sensation to the lips and throat. Water provided no antidote, only bread. Needless to say, we adopted this as a squadron drink.

I returned to RAF Gütersloh in April 1983 in command of 18 Squadron, now equipped with Chinook. We had formed as the first RAF Chinook squadron in February 1982 at RAF Odiham with a move planned to Gütersloh in April 1982. In the event, this relocation was delayed, initially due to a shortage of married family housing. A new purpose-designed hangar had been constructed and fully completed at RAF Gütersloh to accommodate the Chinook squadron and I found it difficult to comprehend why no provision had been made for an increase in family accommodation. Notwithstanding, it was of little consequence for shortly after we deployed as part of the South Atlantic Task Force returning to the UK in August 1982.

A Chinook crew comprised a flight deck crew of captain and co-pilot and two crewman who occupied the rear cabin.

No. 18 Squadron was warmly welcomed back at RAF Gütersloh, a station now full of 'woodland folk' with two Harrier, one Puma and one Chinook squadron all having an off-base operational role in support of 1 (BR) Corps. Our area of operation was largely unchanged from my time in Germany in the 70s and woodland and forest were our anticipated locations for war.

During my time in the Falkland Islands where, post-ceasefire, we operated up to five Chinooks dispersed around the settlement of Port San Carlos, we lived essentially 'off the land'. All our deployment equipment had been lost on the *Atlantic Conveyor* and we lived in a variety of permanent accommodations, ranging from houses to sheep sheds, in Port San Carlos. This was to influence greatly my future thinking on the operation of Chinook

Heading for the South Atlantic on MV *Atlantic Conveyor.*

in the field. Due primarily to the size but also the effect of the rotor downwash, it would prove difficult to hide the aircraft in woodland/forest. I instead favoured light industrial complexes providing good hardstanding and a variety of buildings allowing the aircraft to be dispersed among them. These buildings would also provide for our technical and domestic requirements. However, I was never consulted by those whose responsibility it was to write our 'concept of operations'. Instead, shortly after our arrival at Gütersloh we received a holding of tents, field kitchens and all other equipment deemed necessary for off-base operations. In effect, the scaling for a traditional-support helicopter squadron, albeit slightly increased. I recall questioning the rationale at the time but to no avail.

While in the Falkland Islands we had gained considerable expertise in operating with underslung ISO containers and I saw this as a way forward. Having acquired the necessary twenty-foot ISOs, we converted one to a field kitchen and a further one to a technical facility. On deployment from our main base to the field, Chinook would undersling these to the field location giving an immediate operating capability – a cup of tea. Sadly, headquarters thinking still favoured the lorry-borne deployment of equipment.

Aircrew, having been issued with the AR5 respirator tended to store it on top of their flying locker hoping that the problem might go away. I decided that we needed to work up a capability of operating in this new equipment by both day and night. The single biggest problem was the strict control of decontamination and here I secured the services of our RAF Regiment personnel. We set up a donning and doffing facility under their supervision

Tony Stables meets the Falkland Island natives.

A Chinook at San Carlos waiting to refuel.

and flew all sorties by day and night with one pilot and one crewman from the aircraft crew of four wearing the AR5 with the associated filter unit – all aircraft were modified to stow these 'handbag' units. The donning and doffing processes, done properly, were very time consuming and the equipment certainly slowed us down. However, with practice everything became possible even flying with night-vision goggles. My tour in Gütersloh ended in September 1984 culminating in Exercise Lionheart, billed as the largest land-air exercise in Germany since World War Two.

RAF Germany throughout the period of the Cold War was an exciting place to be. The mission was clear, we trained very hard, and we were very good at what we did. We also enjoyed life to the full. These were unique times, a war without a war, but with hindsight a war that we could very well have won.

CHAPTER EIGHT

TORNADO AT BRÜGGEN

MIKE NAPIER

When I arrived at RAF Brüggen in autumn 1985, the Cold War was still in full swing: the Russians were expected to swarm across the Inner German Border at any moment. Sitting on the Dutch border to the west of Mönchengladbach, RAF Brüggen, bastion of the RAF's front line, exuded a powerful sense of purpose: a fortress of barbed wire, concrete bunkers and drab green buildings lurking menacingly along the edge of a pine forest. In the QRA shelters at the end of the runway, nuclear-armed aircraft were at fifteen minutes' readiness to launch and strike at Warsaw Pact forces in eastern Germany and western Poland.

Mike Napier in his 14 Squadron 'office'.

The runway at Brüggen lay east-to-west with the four squadron dispersals at each corner. Each squadron site comprised nine hardened aircraft shelters (HAS), a supposedly bombproof structure that could accommodate a single Tornado. There was a personnel briefing facility, which was locally known as 'the Hard' and contained all the planning facilities and the operational and engineering controllers who ran the flying programme. Two other long, low huts, which were probably intended as temporary structures when they were built in the 1950s, were known as 'the Soft' – one for the engineers and one for the aircrew. Our one contained the crew room in which off-duty crews could lounge with a cup of coffee and talk about flying.

Once or twice a month, the wail of sirens would rouse me from my slumbers in the early hours of the morning. Still only half-awake, I would don my 'war suit' (a plain flying suit devoid of any unit badges) grab my 'tin helmet' and gas mask and race into the squadron. Once there, the first question was always "Is it real, or exercise?" Often, we wouldn't know the answer for quite a while – meanwhile the briefing slides and mission folders for our pre-planned 'Option Alpha' wartime target were issued and the aircraft were loaded with live bombs. Yes, life at Brüggen was exciting stuff.

Exercises were taken deadly seriously at Brüggen. The station had, over the previous two decades, established a reputation for excellence throughout NATO in the annual Taceval, which tested the entire station's ability to fight a war. There was massive pressure to keep up the 'excellent' scores and not to screw up. As a result, over the next few years I was to

'Goldstars' pose with a Tornado in a hardened aircraft shelter.

spend many hours wearing my 'tin helmet' and gas mask when most sensible people were still tucked up in bed. I also spent numerous weekends manning an aircraft in the QRA shelters. This duty was a twenty-four-hour stint in the QRA area, a cluster of HAS fortified heavily with barbed wire and guarded by armed policemen. After checking into QRA in the morning, the first duty was to take over responsibility for the weapon system from the off-going crew. Although we had practised the drills many times, there was still a moment of awe when you stepped into the HAS and saw the real bomb nestling malevolently under the aircraft. That was the moment when you realised the enormity of what you were doing. Once the formalities were completed, and I had become the temporary owner of an explosive quantity of uranium, there was nothing else to do for twenty-four hours, except watch videos and wait for World War Three to start. After the novelty of doing it once, QRA became a chore, and it was quite usual for the flight commanders to find some reason why one of the junior boys should do their duty for them. As one of the junior boys, I spent lots of time there.

We worked hard, but we played hard, too. The centre of our social life was the officers' mess, another dull green single-storey, and very temporary-looking, building on the north side of the airfield. Here the epicentre was the bar, where the unfeasibly large Driek and matronly Hannah-Lohre served foaming beer at unbelievably cheap prices. The bar was always buzzing and when I first arrived it was de rigueur for everyone on the station to call in there on the way home. It was a wonderful opportunity for us to meet up and discuss the day's business. 'Happy Hour' on a Friday was when the mess bar really came alive, populated by a raucous crowd of officers from across the station consuming prodigious amounts of alcohol … it wasn't politically correct and it probably wasn't sensible, but my goodness it was great fun!

Our bread-and-butter flying was in Germany, nearly all of which was a 500-ft low-flying area, although it also contained some 250-ft low-flying areas. There was no shortage of realistic targets – SAM sites, logistic sites, airfields, deployed army units, bridges to name but a few categories – and the skies were filled with other aircraft, against which we practised our air-to-air tactics. And just in case we didn't have enough trade, a phone call to one of the Phantom squadrons at Wildenrath usually secured some guaranteed opposition. It was not difficult to put together a challenging (and dare I say enjoyable) one-hour forty-five-minute sortie, which included two simulated attacks, some fighter affiliation, and a range detail. The usual range was at Nordhorn, on the Dutch border to the north-east of Arnhem. The target was a bus in the middle of a large circle, which sounds easy enough, but trying to find it on a murky German day from the attack height of 150 ft at 480 kts could be a challenge. Where the IP for UK ranges were huge bridges or lighthouses, the IP at Nordhorn was an anonymous and easily missed minor road junction. From here you headed virtually due

A Tornado GR1 of 14 Squadron.

south, dropping down from 1,500 ft to 150 ft. About a mile short of the target was a line of cottages, which we thought must have been inhabited by deaf Russian spies: no-one else would have put up with being overflown at ultra-low level by NATO jets every few minutes. Luckily the cottages provided a bit of a line-up feature as the target was conveniently masked by trees behind them until the last moment. If Nordhorn range was almost impossible to find, the Belgian range at Helchteren was a much better prospect. Here the main challenge was staying out of the control zone of Maastricht airport, but if one did so, a massive well-marked lead-in took you to a huge yellow board which was almost impossible to miss. There was also a Dutch range at Vliehors, amongst the sand flats on the tip of Vlieland Island, which, though not as easy to find as Helchteren, did benefit from some very generous range safety officers: "OK you missed the target, but I'll give you a direct hit on the offset."

Night flying on the continent was a rather tedious chore: it involved a couple of set routes, the 'Night Charlie' which flew anti-clockwise round northern Germany and routed through the range at Nordhorn, and the 'Night Bene' (or 'Benny') which flew clockwise through Belgium and the Netherlands, routing through Vliehors range. They were all designed for use by non-terrain-following radar (TFR) aircraft droning round at around 2,000 ft and offered very little training value to Tornado crews. They were also flown under radar control, which seemed rather bizarre for supposedly tactical training, but which did offer some amusement on one dark night: tasked with a Night Benny, we were allocated the callsign 'Mission 1111' and we spent the evening trying to see how many ways we could say the digits as we checked in with the different radar units. One-one-one-one, one-treble-one, eleven-eleven, one-double-one-one … well, it kept us amused for one and a half hours.

The UK offered a much better quality of night flying, and we frequently spent our evenings engaged in 'high-low-highs' to make use of the low-flying areas and ranges in the country. At night we could use the TFR and low fly pretty much as we pleased at 500 ft. These night trips were invariably flown as pairs in parallel-track formation. The concept of parallel track was that two individual routes would be planned (one for each aircraft) parallel to each other and two miles apart. As the aircraft flew their own routes, they would be in battle formation with each other. Since we could not fly crossover turns using the TFR, the aircraft could keep roughly abreast of each other by planning minor changes in ground speed and flying large turns as a series of smaller ones. This technique was known, even to navigators who had only known decimal coinage, as 'thrupenny-bitting'.

The range at Jurby, which lay along the west coast of the Isle of Man, was often used on these sorties. At night there was no artificial light there, so you could see nothing outside: it was completely black. Jurby was not a popular range with navigators because, although there was a distinctive 'wine glass' pattern of buoys and target rafts on the radar, it was also very easy to mistake one of the many fishing vessels in the area for the target raft. Invariably we carried out a loft attack there – throwing the bomb forward as the aircraft pitched up at 4 g. It was a surprisingly accurate attack profile, but recovery from the manoeuvre basically

Aircrew of 31 Squadron. Mike Napier seated third from left.

involved flying aerobatics on instruments close to the sea – a particularly scary manoeuvre at the best of times, but even more so when the thing that you might hit is invisible, so Jurby was not a popular range with pilots either. There was also the small matter of the long bum-numbing, bladder-stretching transit home.

Twice a year we headed to Sardinia, once for our annual armament practice camp (APC) and once to use the air combat manoeuvring instrumentation (ACMI) range. Our temporary home for two weeks would be the large airfield at Decimomannu (known colloquially as 'Deci'), which sprawled amongst the dusty olive groves on the central plain of Sardinia, to the north-west of Cagliari. It was run quadri-nationally by the British, Americans, Germans, and Italians, and it frequently also hosted the French, Canadians, Dutch and Belgians, among others. Going to Deci was rather like going on a Mediterranean activity holiday: a busy time flying in the daytime and a relaxing social scene in the evenings. We were quartered in a rather squalid accommodation block, which boasted its very own bar, known to all as the 'Pig & Tapeworm'. The Pig & Tapeworm was just a large room, empty but for two huge fridges which were stocked with cans of beer. It was decorated with murals painted by all the RAF and RN units that had been to Deci and was the scene of many a debriefing; it was also a great place just for letting off steam. Around the corner was the much more up-market Quadri-National Officers' Club, run by the Italians. For some reason it was a tradition when visiting the club for us to drink cappuccino and sambuca, two things we never drank anywhere else. The result, after the first evening's inevitable overindulgence, was a very poor night's rest as you lay on the bed with eyes wide open thanks to all the caffeine, while the room spun madly thanks to all that sambuca. There were usually quite a few bleary, red-rimmed eyes at met brief on the first morning.

Apart from drinking sambuca, the reason for going to APC was to use the excellent range on the west coast of Sardinia. Capo Frasca was one of the few ranges in Europe where virtually every possible weapons delivery profile could be practised. The bombing targets and strafe panels were set out on the top of a headland surrounded by 300 ft cliffs. This had caused problems more than once in the distant past when the occasional Starfighter pilot, used to the sea-level ranges in northern Europe, had let down over the sea to 200 ft on a hazy day and drilled a neat hole just below the top of the cliffs. For us, though, the range at Frasca was a wonderful opportunity to practise all kinds of esoteric methods of dropping bombs and firing guns. We flew in four-ships at Deci, launching in waves to use the thirty-minute range slots allocated to the squadron during each day. With a very full range programme catering for the needs of four nations at a time, there was no time for any flying other than the transit to and from the range and twenty minutes of mayhem on the range itself. The airfield was also continuously busy, as relays of four-ships from all the detachments there departed towards, or recovered from, the range.

Tornados of 31 Squadron head out over the Aegean.

The ACMI range off the west coast of the island was primarily intended for use by the air defenders, but we mud-movers were also allowed on it once a year. At its heart was a telemetry system which picked up the data relayed from a pod carried by each aircraft. In the ground station, all that data produced a real time three-dimensional picture on a huge display screen, showing the positions of all the aircraft on the range and how they were manoeuvring. You could view the range from a bird's-eye view or you could see it from the perspective of the cockpit of any of the aircraft. It seemed like an incredibly complex arcade game, but it was an invaluable tool for working out what exactly happened in a fight involving lots of aircraft, and therefore piecing together a meaningful debrief. Previously, there was enough of a grey area as the various participants tried to put together what they could recall of a combat that the fight was often won or lost during the debriefing. Now, ACMI's 'eye in the sky' ensured that the truth was clear to everyone!

On my first visit to the ACMI we were accompanied by three Harriers from Gütersloh. After starting off by fighting each other, we worked up to two-versus-two combat with them. During one of these sorties my navigator very kindly tried to inspire me by playing the soundtrack to the film *Top Gun* on the cassette tape usually used for loading the route into the main computer. My recollection of these combats was of two very manoeuvrable, but not very fast, Harriers going round in tight circles while the faster but much less agile Tornados slashed past in huge arcs, rather like ice-skating hippos. It was, however, tremendous fun.

Another annual two-week detachment was to Goose Bay in Canada. Located at the southern tip of Lake Melville in Labrador, Goose Bay was a relic from the days when transport aircraft couldn't cross the Atlantic all the way from the USA in one hop. In the late 1980s it enjoyed a reprise when its isolation made it the perfect training area for the RAF and Luftwaffe. The citizens of western Europe were increasingly intolerant of noise from low-flying jets and the vast tracts of uninhabited wilderness around Goose Bay were ideal for practising ultra-low-flying if the weather was good enough or automatic terrain following if it was not. In Europe we were not routinely allowed to use the TFR in cloud in case there were other aircraft around who didn't realise that low clouds and fog might be full of Tornados. At Goose Bay we had exclusive use of the training areas, so we knew that nobody else would be around.

There were two massive training areas, one to the north and one to the south of Goose Bay. Both areas were huge tracts of rolling wooded hills intersected by lakes and meandering river valleys. In the northern area the trees became progressively smaller and smaller as you progressed northwards, until eventually there were none, just bare rock and – even in the summer – ice. The Canadians advised us not to try and judge our height by looking at the trees, because 100 ft over the tall trees of the southern area would equate to about ten feet in the northernmost reaches. By far the best flying to be had in the northern area was Harp Lake, a fifteen-mile finger of water about half a mile wide and surrounded by 2,000 ft sheer cliffs. It was exhilarating to drop into the gorge and skim across the water with the vertical rock faces towering above on both sides, so we tried to do it as often as the plan would allow. Operational low flying, as flying down at 100 ft was officially termed, was thrilling and exciting, but very hard work with disastrous consequences if you got it wrong. The fact that no one did kill themselves doing it in Goose Bay is testament to the excellent training and high standards of our flying skills. There were the occasional close shaves and sometimes aircraft – usually Luftwaffe Phantoms – recovered to Goose with bits of conifers hanging off them.

Deci and Goose Bay were our regular detachments, but the jewel in the crown was undoubtedly Exercise Red Flag, which came around every couple of years. The USAF hosted the exercise at Nellis Air Force Base, on the northern outskirts of Las Vegas, making use of an area about the size of Wales in the middle of the Nevada desert. Here they had assembled an army of Soviet missile systems defending several full-size mock-ups of airfields, railway yards and other realistic targets. The defences were bolstered by a squadron of fighters which operated to Soviet tactics and to cap it all, the whole range was covered by the same telemetry-data system used in the air-combat range at Deci. Taking part in Red Flag was reckoned to be the closest you could get to going to war without doing it. It wasn't without risk, either, and over the years more than one fighter pilot had ended his career by spreading himself across the Nevada desert.

For a Red Flag mission, a 'package' of fifty or sixty aircraft would form up at the eastern end of the ranges ready to push westwards towards the target areas. The strike aircraft would be supported by fighter sweep, 'Wild Weasel' defence-suppression aircraft, airborne jammers, and airborne early-warning aircraft. Getting all that lot together was an incredibly complex business and needed a lot of practice. But, while the planning might have been complicated, the flying was out of this world – every schoolboy's dream. We only flew one sortie a day, usually as a six- or seven-ship formation for one hour and fifteen minutes of adrenaline-fuelled pandemonium. The range work was thrilling, exhilarating, and exciting – and the recovery was often just as frantic as fifty-plus aircraft, all short of fuel, jostled for priority on the way back to Nellis.

CHAPTER NINE

MEDITERRANEAN PATROLLERS

ROD SAAR

In August 1961, No. 79 Entry graduated from the RAF College Cranwell and three of us – Alan Fern, Nick Nixon and me – were posted as navigators to the Maritime Operational Training Unit (MOTU) at RAF Kinloss prior to a posting to a Coastal Command squadron. There were only four navigators on our course (all pilot officers) plus several pilots, flight engineers and signallers. The course started in October 1961 and the first training flight in a Shackleton T.4 took place on 27 October and lasted five hours fifteen minutes. The Avro Shackleton was a derivative of the famous World War Two Avro Lancaster. Apart from lectures, the course consisted of flying exercises using navigation search techniques, sonics to track a submerged submarine, live bombing, and an exercise with the RN submarine *Tireless*. The course finished in January 1962 having completed almost 100 hours airborne training. I had requested an overseas posting and was delighted to be heading for 38 Squadron in Malta.

Rod Saar.

The squadron was based at RAF Luqa on the south side of the airfield, with 39 Squadron (Canberra PR.9s) on the west side and the civil airport on the north side. The station adjoined the airfield, but the officers' mess was about a mile away just off the main road to Valetta. At the time, Malta was still a major Royal Navy base with warships in Grand Harbour. Visible bomb damage from the war could still be

RAF Luqa, Malta.

seen in Valetta. The people of Malta were extremely friendly, speaking a unique language derived from Phoenician, Italian and Arabic, but most also spoke English. It was a strongly Roman Catholic country. In the countryside people still went about bare-footed during the week.

There was a strong contingent in the officers' mess, of very junior RAF officers on their first RAF flying tour. Soon after arriving in Malta, I took a Maltese driving test, as I did not have a full UK driving licence, and then bought a new Mini Minor as, to have any kind of social life, you needed your own car. From May to September the squadron routine, when not flying, was to arrive at 0700 hours and stop work at 1300 hours. Instead of lunch in the mess, a packed lunch was taken to the officers' club beach (rocky with only one sandy beach in the whole island) near Sliema. Being in the middle of the Med the climate was very mild in the winter and in mid-summer could be very hot. It was a great place for a posting.

On arrival in February 1962, I joined Flight Lieutenant Paddy Shorthouse's crew as second navigator (our co-pilot was Flight Lieutenant Chris Taylor, who died when an RAF Kinloss Shackleton crashed into the Moray Firth in 1965). Flight Lieutenant Ted

Pike, a World War Two veteran, was the first navigator. The squadron had eight crews and each consisted of two pilots, a flight engineer, two navigators, an electronics officer and four NCO signallers (one W/T operator, one radar operator and two port and starboard lookouts manning an F.24 handheld camera). On the squadron all the crew captains were pilots, but on other RAF maritime squadrons they also had navigators as captains. The squadron was commanded by a pilot, Wing Commander Harding, later replaced on posting in 1963 by Wing Commander Thomas, a navigator. There were two flight commanders, Squadron Leader Mo Short and Squadron Leader Bowyer, both pilots. There was also a navigation leader, Flight Lieutenant Bob Davies, who monitored the progress of the navigators, particularly the first tourists, and an electronics leader responsible for the standards of the electronics/signallers. The squadron had its own airman ground crew who carried out first-line servicing. It was good for their morale to be members of a squadron, and from time to time they were taken on detachment with an individual crew. Sadly, for reasons of economy, the RAF later centralised all first-line servicing, so the airman lost that sense of identity and the close relationship with the aircrew.

My first flight as second navigator took place on 23 February 1962 on a fourteen-hour NATO Exercise, Casex 73, off Sicily, hunting submarines in cooperation with NATO navies. On exercise we carried 8.5-lb break-up bombs to simulate depth-charge attacks on submarines. Patrols were flown at 1,000 ft (so we never carried parachutes) and attacks were made at 100 ft by day and 300 ft by night, so the radio altimeter was a vital piece of kit. The aircraft had a wingspan of 120 ft, and a sharp turn following an attack put the aircraft's wingtip very close to the sea.

Shackleton MR.2 of 38 Squadron.

Navigation in the Med was back to basics as there were no aids such as Gee or Loran. Later, we had Green Satin installed, a ground-position indicator but, with an invariably calm sea, it unlocked due to a poor Doppler return signal. Over the sea at low level, D/F radio bearings were also very limited. The API (air position indicator –the aircraft's ground position assuming no wind) was a critical navigation aid, followed by the handheld sextant and the drift meter fitted in the tail lookout position. Fortunately, at low level, experienced pilots were able to accurately estimate wind direction and speed from the 'sea caps'. One navigator had the role of 'routine navigator' doing the geographical navigation assisted by the 'tactical navigator' (they sat side-by-side) who carried out the astro calculations when the sextant was being used by day for sun position lines and by night for a three-stars astro fix. The 'tactical navigator' sat at the plotting table and tracked a submerged submarine following the launch of a pattern of three sonobuoys under his direction. In an attack the 'routine navigator' became the bomb aimer lying flat in the nose position and directing the pilot onto the target before releasing the weapon.

The main aid to detection of submarine snorts or surface vessels was the ASV-21 radar and the 'Mk 1 eyeball'. The ASV-21 was manned by a signaller on thirty-minute shifts in a totally dark area curtained off with blinds. From 1,000 ft, a ship could be detected at forty nm whilst the snort of a submarine at max ten nm. But the sea state was key to detecting a submarine as in a high sea state the submarine snort disappeared in the radar clutter caused by high waves. Once a snort had been detected, the radar operator gave the captain a range and bearing before switching off to minimise the possibility of the submarine's D/F alerting them to the presence of an aircraft. The captain would call 'action stations, action stations, go' alerting the whole crew, the routine navigator hurrying down to the bomb aimer's position. The nose position would be manned by a signaller to look for a visual contact of the submarine snort. Towards the end of my tour, we were equipped with Orange Harvest, a passive radar detection receiver. Another piece of detection kit was Autolycus, a device which detected any fumes generated by a submarine. We tried it once on exercise and found it useless because ships also gave off fumes. It might have been more useful in a very large sea area, such as the Atlantic and outside shipping lanes.

The squadron's secondary role was search and rescue (SAR), and each week a crew spent one day on SAR standby. The primary aim was to locate a ship in distress, or a dinghy or someone lost in the Libyan Desert. Once a small target was detected, the position was marked with a flare and a call for shipping assistance was made with the aircraft remaining on station. For a major incident, we could drop a Lindholme Gear, which consisted of an MS.9 dinghy which deployed on hitting the sea. Attached to it were several floating containers containing survival equipment. In my time we only did one exercise launch of the Lindholme Gear close to Malta with an RAF launch in attendance. I recall only two live call-outs when on SAR standby. The first occurred on 27 August 1962 when a seaman in the rear seat of an RN Meteor ejected (possibly accidentally) near the coast not far from Grand Harbour. We were airborne at 1215 and spent over seven hours searching until dusk, but without success. On another occasion, we were called out to search for a lone oilman travelling in a vehicle who was reported well overdue in the Libyan Desert. The desert is not flat but has undulating dunes with some mountain ranges further south. Looking for a lone vehicle in the desert is like a 'needle in a haystack'. Surprisingly, at night across the desert we saw many very small flickering lights indicating the presence of Bedouins. In our search we flew at 1,500 ft, but the desert haze made detection of a small object very difficult with very limited information on a likely route taken by the oilman. After many fruitless hours we returned to base and the following day learnt that the oilman had eventually turned up at base. On 26 October 1963, Crew 7 (the second crew I joined on 30 November 1962 as second navigator), with Flying Officer Mike Beane as captain, were tasked with an SAR escort role to cover the transit of three RAF Belvedere helicopters flying from Brindisi to Suda and on to El Adem, with most of the route over the sea. We took off at 0500 hours and spent over eleven hours in the air, half the time escorting the Belvederes safely to El Adem where we landed at 1620 flying back to RAF Luqa the following morning.

Low pass of Marble Arch in Libya.

In addition to maritime operations and SAR duties we did some long trips to exercise navigation skills. On sorties back to the UK to RAF St Mawgan and RAF Kinloss we usually flew airways. But the most demanding trips were from Malta to Aden and return; I did two of these trips during my tour. On 16 March 1962 we did the short first leg to El Adem, landing around lunchtime. We refuelled and rested in the afternoon for a 2100-hour take-off as we planned to use astro navigation throughout the night. Our track took us close to the south-western edge of Egypt known as 'Nasser's Corner' which, having a rocky feature, was picked up on the ASV-21 radar giving us a fix. The first navigator and I took turns on a three-hour basis to be the routine navigator taking half-hourly three-position line sextant star shots whilst the other navigator did all the calculations to plot the fix. We flew at 9,000 ft, without oxygen, and through the night travelled across the Sudan flying over Ethiopia by morning. We could now clearly map read, although some features did not appear correct on our maps. We landed at Embakasi, Nairobi after a twelve-hour thirty-five-minute flight. Apart from a trip to a game reserve and purchases of wooden carvings, the captain arranged for the purchase of 600 pineapples which were loaded into the two bomb-bay panniers. They were bought for one shilling a piece and on return to Malta were sold for two shillings with all profits going to the squadron's airman's fund used to help any airman in financial difficulty.

We left Nairobi on 19 March at 0830 using a mix of aids for navigation, some D/F and sun-shot position lines and arrived at RAF Khormaksar after six hours. We were accommodated in a rather basic hotel in the city with a smelly goat farm visible behind the hotel. Transport Command crews enjoyed a much more luxurious hotel despite a more comfortable flight than in the very noisy Shackleton. At this stage, Aden was free of troubles, so we shopped in the centre of town. Malta was not a 'duty-free area' so colleagues and service friends in Malta placed their orders for primarily electronic kit e.g., tape recorders etc. This was stored in the nose and the rear of the aircraft, and with our pineapple load we were very close to our maximum all-up weight (AUW). Our major concern was not to be delayed in Aden since, in the heat of the day, the pineapples would start to rot, and we could smell them from 100 yards. In view of the daytime heat (length of runway needed for take-off close to our AUW), our intent once more was to use mainly astro navigation on our return to Malta, so an evening flight was planned. We took off at 1945 hours for a direct flight back to Malta landing thirteen hours later. So, in five days, we spent almost thirty-five hours in the air with a lot of good astro-navigation practice and an enjoyable change from our normal squadron routine.

Anti-submarine training was a constant feature for squadron crews. On 9 July we flew to RAF Ballykelly to attend the three-week Joint Anti-Submarine School (JASS) course focused on cooperation with the RN in anti-submarine operations. We were involved in six separate exercises giving us a total of fifty hours flying time. On return from one exercise we came across an RN nuclear submarine on the surface returning to the Clyde, which left a very distinctive wake. Back in Malta, we regularly practised bombing, using the 8.5-lb break-up bombs against towed targets by an RAF launch or a radar buoy by night which also gave the ASV-21 radar operators target acquisition practice. We did one live depth-charge practice attack and one actual launch of a Mk 30 torpedo, including launching a pattern of the three acoustic sonobuoys.

One other aspect of training was for the whole crew to be taken out by RAF launch out of sight of Malta before being dropped into the sea and clambering into a large dinghy. Some hours later, we were located by one of the squadron's Shackletons and recovered by the RAF launch. What was very strange was that although a large distance from the Libyan coast or Malta we were suddenly pestered by flies. Prevailing winds often came from the Libyan Desert and in the summer the 'Sirocco wind' would bring a heatwave to Malta.

NATO exercises with Royal Navy and other navies was a constant preoccupation for the squadron when NATO submarines played the role of the enemy. Most of these exercises took place in seas south of the toe of Italy to the east of Sicily and were designated 'casex'. I flew on nine separate casex exercises plus several other exercises with the RN and a couple with the British Army over the Libyan Desert, one involving eleven hours in support of the Royal Fusiliers. Most exercises were flights of over ten hours. On 11 June we took part in a casex over the Aegean Sea hunting Turkish submarines. During our ten-hour sortie, we detected one submarine which dived before we could launch an attack. However, having noted the position we left the area and returned an hour later with our ASV-21 off and found the submarine snorting allowing us to make a successful attack.

A Royal Navy S-class submarine

With Crew 7, the most successful exercise was Exercise Moontiger in April 1963 with the Greek navy off Crete when we recorded three successful submarine kills during a fourteen-hour sortie; each attack was photographically recorded.

Every year we flew to RAF Gibraltar for an exercise with the Spanish navy and the RN. In November 1963, we carried out three sorties on Exercise Moontiger II, and on the last sortie we recorded a kill against HMS *Turpin.* On one of the 'kills' I was the bomb aimer when one of the two 8.5-lb break-up bombs hit the submarine periscope forcing it to surface and return to base; I was rated as an above average bomb aimer – it was great fun.

In June, we flew to Mombasa via El Adem and Nairobi for Operation Pipe. The Soviets had satellite-tracking ships in the Indian Ocean and our task was to locate one of them and photograph the tracking gear if displayed. We were given a large area to search. We carried out two sorties of fifteen and twelve hours respectively. Navigation was again very basic because we could not use Green Satin since the Doppler signals could be intercepted by a Soviet ship and we were too far out to sea to obtain any D/F radio bearings. We used the sextant to get single-position line sun shots and used the drift sight, having dropped a flare float and taken drifts on three separate aircraft headings, allowing us to compute a wind. Using the API, which tracked the aircraft ground position assuming nil wind, the drift meter wind and the single-sun position line we worked out our MPP (most probable position). I recall an element of satisfaction when after ten hours flying, using these very basic navigation tools, on arrival at the Mombasa coast the ASV-21 gave us a radar fix and we were five miles away from our dead-reckoning position. During our two sorties we came across one Soviet tracking ship, but the gear was not up. During the thirteen days away from Malta we had flown almost thirty-two hours. No. 37 Squadron from Aden took over the task and caught a Soviet satellite-tracking ship with its gear up. During 1962–1964 there was only one occasion in the Med when Soviet navy ships were in transit, so unlike the UK squadrons, we had little experience of tracking the Soviet navy.

My final months with the squadron were taken up patrolling the Aegean Sea north of Cyprus following the outbreak of hostilities between the Greek and Turkish Cypriots, which led to the island being split. Our task was to monitor any naval activity by either the Greeks or Turks, which at that time, the end of 1963 into 1964, might indicate an imminent invasion by either party. On 27 December 1963, we were on SAR standby and in the afternoon were called to operations and briefed to fly to Cyprus and patrol the Aegean Sea to check for any naval activity. After thirteen hours we landed in the early hours at RAF Akrotiri in Cyprus. From this point, the squadron had two aircraft permanently based in Cyprus, mainly RAF Akrotiri, but occasionally RAF Nicosia, carrying out twenty-four-hour coverage of the Aegean Sea. Most sorties lasted just over twelve hours before we handed over to 37 Squadron aircraft. So, for three weeks we flew on alternate days before going back to Malta for a week's rest before returning to Cyprus. At the start of January 1964, I changed crews (captain Flight Lieutenant Russell) when I became the first navigator.

There were two events during this time in Cyprus which are worth mentioning. Early on one of our night sorties we became aware through intercepted R/T communications that another aircraft was in our airspace. We were flying at 2,000 feet without lights. It turned out that the Group Captain Operations at RAF Episkopi had tasked an RAF Hastings to carry out the same ship recce role as ourselves in the same area at a level just above us and also without lights; neither crew had been briefed of the presence of the other aircraft. The Hasting crew were relying on visual sighting whilst our ASV-21 radar could detect shipping up to forty nm away. Representations were made by our captains to the group captain, a bomber man, who had no idea of the operational capability of a Shackleton, but he refused to listen. In the end, an air commodore from the MoD was sent out to Cyprus to sort the matter out. This illustrates the occasional tendency I witnessed in my twenty years in the RAF of arrogance by virtue of rank. Two aircraft could have suddenly disappeared at night in the Aegean Sea due to this ignorance which could have cost lives.

The other event was unusual and surprising. On one of our sorties to monitor naval activity in the Aegean Sea, we had flown early one morning at very low level into the Bay of Iskenderun to check on the Turkish navy in the harbour and to take photographs. A couple of weeks later, we repeated an early morning harbour penetration and, as we left the area, we were aggressively buzzed by two Turkish F-86 fighters, one opening fire to scare us off. The port beam lookout took photos of the incident with the handheld F.24 camera. We descended to 100 feet above the sea and returned to base; the F-86s did not pursue us. Back at RAF Akrotiri the captain reported the incident, which was passed to HQ at Episkopi. The initial reaction was that we were mistaken since Turkey was a member of NATO. Once we produced the F.24 camera photos of the incident, the matter went quiet, but of course we had gone into Turkish airspace without clearance. They had clearly not intended to shoot us down, but merely scare us off – they succeeded.

From 27 December 1963 to 31 July 1964, I completed thirty-two Operation Survex sorties consisting of 173 hours day flying and 145 hours at night. Some fifty-one years later, to my great surprise, I was informed I could apply for a General Service Medal (with Cyprus clasp) in recognition of service in Cyprus during the 1963–64 emergency. A bit strange when considering that more risky and stressful RAF operations on the ground in East Germany during the Cold War have been turned down for a GSM by the MoD no less than three times.

Finally, it is worth mentioning the hazards faced during Shackleton maritime operations due to the low-level role, often in very tough weather conditions far out to sea. In the history

a mess with detritus from the galley and other personal items, and there was total silence all round. We were then diverted to Kenitra, which was a USAF base in Morocco. As we taxied in, a group of USAF personnel were waiting armed with cameras and could hardly believe that this four-engine 'World War Two lookalike' was still flying operationally. They were even more surprised when fifteen people emerged from the aircraft. We discovered we had lost our tailwheel doors and several engine exhaust stubs.

On 27 May, having recently returned from RAF Nicosia after three weeks on Operation Survex, we were tasked to carry out a live depth-charge drop. I was first navigator and bomb aimer on Squadron Leader Barrett's crew. The drop was against a fixed-radar buoy some way out to sea. We took off at 1730. Following the drop, the tail lookout reported seeing only one depth charge hit the water. I checked looking through the bomb-bay inspection panel to discover the other depth charge had hung up. We immediately set course for home where the captain carried out a very smooth landing. As we got into the dispersal, there was a sudden loud thump as the depth charge fell off its mounting onto the bomb-bay doors. Before the captain had time to say, "abandon aircraft", we were already hot footing it well away from the aircraft. In theory the depth charge would not explode because its detonator was set against a water pressure of around twenty-five feet, but we were taking no chances.

My last flight with the squadron was on 22 August 1964 returning from Akrotiri. During the tour I had flown 221 sorties amounting to 1,540 hours and twenty minutes (thirty-one per cent in the night) flying time. For me this was a very memorable tour based on a pleasant and friendly island in the Med. The role was very varied and working in a ten-man crew, where very close teamwork was essential, was very rewarding. On my return to the UK in September 1964, I drove back from Malta (three ferries) in my Triumph Spitfire, stopping off in Frankfurt-am-Main to visit my German paternal grandparents, then on to Brussels to stay with my Belgian step-grandparents. In November 1964, I did a short conversion course at RAF Bassingbourn on the Canberra (very glad I did not do a tour on this aircraft strapped onto an ejector seat in the back in dark cramped conditions), followed by the Staff Navigator Course at RAF Manby and then posted as a navigation instructor to the RAF College Cranwell; my five flight cadet navigator students on 90 Entry spent a lot of flying training time over the North Sea, but all successfully got their navigator brevets in 1967. I had come full circle.

CHAPTER TEN

'SHINY SIX' IN CYPRUS

MICK WHYBRO

After navigation training, even though I thoroughly enjoyed keeping immaculate manual air plots and successfully employing astro three-position line fixing to reach my destination at the right time, and in the right place, I far preferred that phase of the course 'flashing' around East Anglia in the Meteor NF14 – my sort of flying! So, I expressed a posting preference for night fighters. Unfortunately, the Javelin OCU had just stopped taking any more students at this time, so my ambitions were thwarted. And so it was that I was posted to the Canberra ground-attack force – a move which was to set me on the road for what turned out to be a most satisfying eventual career.

Mick Whybro back from a weekend away.

At that time, all Canberra light-bomber squadrons were based overseas. There were three or four outfits in Germany, one in the Far East and four in NEAF (Near East Air Force). The Middle East Air Force was formed in 1956 when the UK offered to provide a light bomber wing to the Central Treaty Organisation (CENTO). This became NEAF with the squadrons based at Akrotiri. The squadrons were 6, 32, 73 and 249. I was fortunate enough to be posted to 'Shiny' 6 Squadron, by far the best of

A Canberra B.16 of 6 Squadron at RAF Luqa, Malta.

the bunch at Akrotiri. This location was looked upon as the most desirable – we ventured as far north as the UK, the length and breadth of the Mediterranean, North Africa, most of the Middle East, as far east as Singapore and as far south as Salisbury (modern-day Harare).

Nos. 6 and 249 Squadrons were equipped with B.16s while 32 and 73 Squadrons had the B.15. The aircraft were identical apart from one big disadvantage in that the B.16s were equipped with a totally useless bit of kit called Blue Shadow sideways-looking radar. To accommodate the electronics for this 'encumbrance' the third ejection seat was removed. This meant that the nav/observer had to make do with a fold-down Rumbold seat alongside the pilot. With no ejection seat, the observer relied on a flying suit which had an integral harness and a chest parachute that clipped onto two hooks on the suit. His escape from the aircraft was via the entrance door which could be opened from the inside by turning a handle at least three-and-a-half turns clockwise. Apart from the missing ejection seat, the navigation table had to be folded up to check that the system was working properly, which it hardly ever was. The radar output was burnt on to specially impregnated paper that had a nasty habit of burning and producing clouds of smoke, which was most off-putting. The system only looked to starboard to a range of about sixty miles. This meant that we were very familiar with the southern coast of Crete when we were heading west.

Apart from the Blue Shadow, the rest of the plotter's back-seat kit was pretty good and reliable. This consisted of Blue Silk Doppler, a ground-position indicator (Mk 4A), Rebecca/Eureka range and bearing equipment, a Marconi radio compass and an STR18 HF radio (complete with Morse key) which was vital for receiving the go/no go message for the nuclear role. There was also an Orange Putter tail-warning radar, but by the time a threat was illuminated, it was too late.

The bulk of our flying was lots of desert low levels and weaponry in North Africa between El Adem (Tobruk) and Idris, Tripoli. Tripoli was a good run ashore in those days. At the time, the UK was a member of CENTO. The stated aim of the organisation was to contain the USSR by having a line of strong states along the Soviet Union's south-western frontier. The other member states were Iran (Persia at the time), Pakistan, Turkey and Iraq, although Iraq was something of a non-participating state. As such, we took part in several associated exercises a year, often based in Tehran for days at a time. It was a fabulous city in those days, very westernised, a real Paris of the Middle East. We also conducted APCs twice a year in Sharjah (a member of the United Arab Emirates since 1971).

We practised five modes of weaponry. The boring one was a requirement to keep current in medium-level bombing straight and level above 10,000 ft. The nav/observer came into his own here doing his "left, left, steady, bomb gone" routine. The training for this was carried out using the local Episkopi bombing range flying numerous circuits in an academic pattern for a couple of hours dropping several 25-lb practice bombs. Far more entertaining and adventurous was when the whole wing was involved. Six aircraft from each of the four squadrons, each aircraft carrying four 1,000-lb bombs, would be launched at night at two-minute intervals. The route would take us south of Crete at around 30,000 ft, before turning left and descending to our bombing height of around 10,000 ft to deliver the stick onto the El Adem targets in Libya before climbing back to a suitable transit height for the return to Akrotiri, all-in-all almost a three-hour sortie. It always amazed me how we managed (most of us anyway) to arrive at the target in the right order and more-or-less on time – says a lot for the skill of the nav/plotters.

The other, far more enjoyable, modes of weaponry were shallow-dive bombing in a thirty-degree dive, two-inch shallow-dive rocketry, level rocketry (quite hairy with a fifty-feet run-in) and finally LABS. This was by far our most important delivery mode, which reminded us that we were key players in the Cold War conflict. It was the method that we would use to deliver our 2,000-lb Red Beard nuclear weapon. In November 1961 a permanent facility, the supplementary storage area at nearby Cape Gata, became available to hold thirty-two Red Beards. Every Saturday morning was spent doing intense study of our intended target. Without going into details, suffice to say that each of the Akrotiri Strike Wing crews had a target in south-western USSR. That was a lot of Saturdays for three years, probably why I can remember the details of our Rostov target to this day.

We also had to brace ourselves for regular visits from the UK RAF 'nuclear procedures checking team' – no errors were tolerated. The fissile core did not come with the weapon and had to be loaded by the aircrew, a procedure known a LML (last-minute loading). The core came in a dustbin-like container from which it had to be gingerly removed before being placed into the weapon. Thank goodness that my job, as the nav/ plotter, was only to read from the checklist; the poor old nav/observer had to carry this manoeuvre out, which could sometimes be tricky, especially with the well-used training kit.

To be honest, the Cold War did not really manifest itself as far as our daily lives were concerned, but it certainly got you thinking on Saturday mornings, and when the UK weapons standardisation team pitched up.

Akrotiri was the home of a 'copybook' tactical air force. As well as the four light-bomber squadrons, there was one fighter squadron (Javelins during my time, followed by Lightnings), a photo-recce squadron (Canberra PR.9s), a transport squadron (Hastings followed by Argosies) and a supporting search and rescue helicopter flight. During my three years there was only one fatal accident, sadly it was from my squadron. On 16 July 1964, WJ771 disintegrated fifteen miles south of Khartoum after entering an active cumulonimbus cloud at 40,000 ft. All three members of the crew lost their lives.

From time to time, we would find ourselves being employed on non-routine exercises. In late 1964, 29 Squadron Javelin formation aerobatic team were tasked with displaying at Shiraz and Esfahan in Iran. This required them to transit through Diyarbakir (central Turkey) and Tehran. The air staff at NEAF decided that because of the unreliability of the weather forecasts in this part of the world, and the lack of suitable diversions at the destination airfield, the Javelins with their limited fuel should be led by a Canberra crewed

The Whybro crew from left to right: Russ Russel, Mick Whybro and pilot Johnnie Wools.

by yours truly. This did not go down too well with the Javelin team leader (also their squadron boss) and so a compromise was agreed upon. We would fly the route ahead of them, do a complete let-down to the destination airfield and then meet the Javelins at their point of no return and give them the gen. In the event this system worked very well. We all stopped overnight in Tehran outbound and inbound, which proved to be very hard work keeping up with the eight Javelin aircrew. They were a good bunch though, all very enjoyable.

Our crew found ourselves doing a similar exercise in February 1965. (We clearly did a good job with the Javelins.) This time, UK Hunters were transiting between Nicosia and Khormaksar, through Diyarbakir, Tehran and Sharjah, and that all worked well. (Although the Hunter guys were not as friendly as the Javelin aircrew – no navs involved.) Also in that same February, our crew were tasked to take spares to an aircraft stuck at Gan. In the event, we went u/s there and stayed for a week. For the transit from Masirah to Gan, a four-hour transit in a straight line looking for the Gan non-directional beacon, I drew a Mk 9 bubble sextant from stores and the relevant tables (the resident Hastings squadron used them), and passed the time taking sun shots, some producing quite reasonable position lines considering that I had no canopy refraction tables. All in all, that month, I completed forty-seven hours flying.

Another enjoyable interlude occurred in early 1966. Major work was required on the runway at Akrotiri, so the squadrons were dispersed for the period. Some went up the road to Nicosia, one unlucky lot went to El Adem while 6 Squadron went to Malta. We used the Royal Naval Air Station at Halfar (HMS Falcon) on the southern tip of the

Air and ground crews of 6 Squadron.

island. At the time, Halfar was the home of the RN Meteor target facilities squadron and the Naval Observers School. It made a pleasant change being introduced to life in a naval wardroom. I also managed to scrounge a couple of Meteor trips. All in all, we were there for six weeks.

The major detachment during my time with 6 Squadron occurred between 10 August and 22 October 1965. As part of the UK's response to the Indonesian Confrontation, RAF Germany B(I)8 Canberras were deployed to the Far East at two-month intervals. These were eventually replaced by NEAF squadron aircraft. The earlier detachments were based at Tengah in Singapore, but then moved to Kuantan some 180 miles north on the peninsula. We were accommodated in two-man tents. The only permanent building on the airfield was the air traffic control tower. Operating from the 6,000 ft runway surrounded by enormous trees – the only lighting being oil goosenecks – sorted the men from the boys when night flying. (I think I preferred night flying from the aircraft carrier flight deck in a subsequent life.) Despite the sweaty, humid conditions life was quite enjoyable and entertaining. The replacement squadron from Akrotiri arrived one crew short due to sickness, so our crew was 'lumbered' into staying an extra month when we were rewarded with a long weekend in Hong Kong – a good tick.

All in all, this first tour on 6 Squadron was a fantastic and most enjoyable time – a hard act to follow. The Canberra force was the breeding ground for navigators to move on to the dare I say, dreaded V Force. So, I was more than delighted to find myself avoiding this fate by being selected for loan service with the Royal Navy, flying the best low-level strike/ground-attack aircraft in the world – the Buccaneer.

CHAPTER ELEVEN

NEAR EAST V-BOMBER FORCE

NIGEL BALDWIN

THE RAF's NUCLEAR-ARMED MEDIUM BOMBER Force was stood down from QRA in the United Kingdom at midnight on 30 June 1969 as the Royal Navy's Polaris fleet took on the responsibility for the national deterrent (which the Polaris successor submarine *Trident* continues to this day). By mid-June that year, the RAFs' Vulcan Force was down to just eight squadrons, the Valiants having been grounded in 1965. No. 12 Squadron, with its Vulcans, was disbanded in 1967 (to free up the squadron number for the first F-111 squadron); by then, the remaining Victor Mk 1s

Nigel Baldwin (centre) and his 35 Squadron crew at Akrotiri.

had been converted to air-to-air refuelling tankers, and the two Victor Mk 2 squadrons had gone the same way.

So, what to do with the eight Vulcan B Mk 2 squadrons? Perhaps I can best cover the eventual answer to this question by describing my own experiences.

In the summer of 1969, I was halfway through my two-year tour at HQ 11 Group at RAF Bentley Priory as the ADC to the AOC. I had spent the previous five years on the Vulcan B.2 straight out of training on the Jet Provost from the RAF College at Cranwell – where we were awarded our wings – and then a four-month advanced course at No. 4 FTS RAF Valley where we had expected to be the first course on the brand new Folland Gnat. But, when we got there the Gnats had not arrived, so we flew the Vampire T.11 – an experience I have never regretted.

Immediately after Valley, I attended the three-month OCU course at RAF Finningley on the also new, all-white, Vulcan B.2. I was then posted, as a pilot officer co-pilot, onto IX Squadron at RAF Coningsby alongside 12 and 35 Squadrons. By this time, most Vulcans were emerging from the factory at Woodford equipped with the 300 series Olympus engine (some 2,000 lbs of static thrust from each engine). After two years as a co-pilot, I got the much-coveted captaincy following the path set by other young Old Cranwellians like John Willis (eventually Air Chief Marshal Sir John – Vice Chief of the Defence Staff), Air Vice-Marshal Tony Woodford, and Group Captain Ray Humphrey. After the usual four-month OCU course at RAF Finningley, I then spent three years as a captain on 35 Squadron at RAF Cottesmore. At the end of that tour, I was expecting to go to Fort Worth in Texas to

A well-camouflaged Vulcan XM646 over the Kourion Amphitheatre, Cyprus.

begin to get to grips with the F-111 but that programme was cancelled (the MOD budget was in crisis) and I was left dangling for three months before getting the ADC slot at Bentley Priory. The air vice-marshal I worked for came in useful and, having kept my nose relatively clean (and his too), in 1970 I returned to the Vulcan B.2 OCU, trained (another four months) with a new crew, and, to our delight, found ourselves posted to 35 Squadron at RAF Akrotiri in Cyprus. Soon after arrival, I was promoted to squadron leader and became one of the two flight commanders on the squadron.

The only V-Bombers to be ever permanently based overseas were the Vulcans of IX and 35 Squadrons at Akrotiri for the first half of the 1970s. There we were a bit limited in scope. The normal training sortie was a high-level navigation stage to Crete and back (Flight Levels 450 to 500) then once around the island anti-clockwise at 500 ft AGL attacking, with a F95 camera, two or three white-painted village churches on the way and then several low-level patterns at RAF Episkopi bombing range dropping up to twelve 28-lb practice bombs on a floating target in Episkopi Bay before back to the Akrotiri circuit for some pilot continuation training. For diversion airfields, we usually used RAF Nicosia in the middle of the island and Adana in southern Turkey.

Another opportunity for different low-level training emerged when we were able to use RAF Luqa in Malta and RAF El Adem in pre-Gaddafi Libya to fly low-level routes up through Italy, usually the western side, and also across the North African desert before dropping practice bombs on the El Adem range. On the face of it, the terrain may have appeared to have had little application to the primary operational mission but remember that, in war, the pilots would have been flying blind (you may have heard of the screen we would have put up to prevent us being blinded by nuclear flashes) and the eyepatches we had in our flying suits and, to the rear crew (who could not see out anyway) the relatively flat and featureless desert, providing few unambiguous, discreet radar returns, was a reasonable facsimile of the western USSR. (I used to pontificate to anyone who would listen at Staff College that the highest ground between HQ Strike Command at High Wycombe and Moscow was Harrow-on-the-Hill. So why were we flying through, and sometimes into, the Welsh mountains?)

Ostensibly, the Vulcan Force was at RAF Akrotiri in both the nuclear and conventional roles as part of CENTO. For those of us at RAF Akrotiri in the early 1970s, times were good. I bought my first Alfa Romeo – a beautiful all-white 1750 Berlina saloon with a wooden steering wheel – for £1,200 duty-free. Perhaps more importantly, I remember a very positive visit by the new prime minister, Edward Heath, (his predecessor Harold Wilson never seemed to have visited the RAF anywhere) and the CENTO Alliance, although a bit suspect politically, gave us a good excuse to range over the Middle East. There were constant detachments of two to four Vulcans to RAF Luqa Malta for low-level and simplex TFR training around North Africa and Italy. There were similar detachments to Tehran (for the same around the Shah's Persia), and to RAF Masirah off the coast of Oman. There were even visits to Mauritius and New Zealand for my crew to display the aircraft.

RAF Akrotiri, with its air commodore station commander – John Stacey (later Air Marshal) and David Craig (later MRAF The Lord Craig) in our time – its Vulcan Bomber Wing operating quite independently of Strike Command in the UK, its Lightnings of 56 Squadron, Hercules of 70 Squadron, Wessex helicopters, RAF Regiment Wing and Bloodhound SAMs, was the RAF's largest base at the time, and was an exciting place, even if all the aircraft were left out in neat packages on the dispersals at night despite Colonel Gaddafi up the road with his Soviet Badger bombers. There was real synergy at RAF Akrotiri, not least because, certainly for the first time in my experience, just about all the elements of air power were together on one station. We were all able to take advantage of our distance – and that of HQ Near East Air Force's – distinct independence from, we thought, the more conservative Strike Command in the United Kingdom.

For our nuclear commitment to CENTO, we did not have seven-day-a-week twenty-four-hour QRA responsibilities, but we did have the same number of no-notice alert and nuclear generation exercises (initiated by HQ NEAF at Episkopi) as our colleagues in No. 1 Group back in Lincolnshire – sometimes instigated by MoD and sometimes by HQ

Nigel Baldwin's crew deliver the Christmas mail to RAF Masirah.

NEAF. Our WE.177 lay-down thermo-nuclear bombs were stored at RAF Akrotiri and were wheeled out and attached to the aircraft regularly (not to be flown in peacetime of course). We were regularly examined by the no-notice 'trappers', a nuclear standardisation unit from the UK. Our station commander had the same generation targets as his No. 1 Group colleagues with a certain number of aircraft at Readiness 15 in a specific time frame – and woe betide if we didn't make it. We usually 'flew off' – but never with live nuclear weapons – in a mass scramble of sixteen or so aircraft at thirty-second intervals after a few days of tension, created by imaginative geo-political scenarios, into a rapidly planned sortie attacking three or four targets with the F95 camera at low level. As we all got back to Akrotiri at about the same time, great was the fun to be had sorting out the landing profile. On one never-to-be-forgotten occasion, the station commander, Air Commodore John Stacey, walked into the control tower as we were all recovering and told the air traffic controller and his assistant to pretend to be dead; and thus, not respond to any of our calls. So, we sorted the landing sequence out ourselves admittedly with the encouragement of a green Very from the corporal (who hadn't been told he was dead) in the caravan at the threshold of the runway. On one occasion, just as we had taxied in and shut down and were planning an excursion to the ski club on Ladies Mile beach, the aforesaid station commander ordered his own 'conventionally armed' (that is, twenty-one 1,000-lb bombs) exercise with a low-level route around the island to be flown clockwise rather than the conventional anti-clockwise route coasting in at Ayia Napa beach, up the pan handle, west along Kyrenia beach in the north and running into Episkopi and Akrotiri from the western tip of Cyprus. And we missed most of the targets – which, of course, was his intention because he knew we were over familiar with the route and too blasé. I remember my own crew's final target, at 300 ft AGL with the F95 camera was the Akrotiri base hospital.

I mentioned our conventional role: the Vulcan had always been a big bombload carrier – twenty-one 1,000-lb HE bombs – by now with retarded tails so that they could be dropped at low level. The black box the navigator radar used, positioned by his left shoulder, was aptly named 'The 90-Way' thus giving us several options. With many 28-lb practice bombs in the bomb bay (all having the same electrical characteristics as their much larger and nastier brothers) we could go round and round the bombing range for an hour or so – unless, as happened to my crew on one occasion, his finger slipped and all twenty-one dropped at once before we had got properly started. We were on detachment to RAF Tengah, Singapore at the time, and our faux pas gave him and me an uncomfortable, and brief, interview with the wing commander operations, the late Air Chief Marshal Sir Michael Knight, then a wing commander. Our little bombs had dropped into the range (fortunately) called China Rock.

By this stage, Taceval had spread its wings from RAF Germany, where our Canberra and then Buccaneer and Harrier colleagues had led the way, into the V-Force squadrons in the

Nigel Baldwin commanded the farewell parade for the Air Officer Commander-in-Chief.

UK and then rapidly to us in Cyprus. The Strike Command Taceval team, led by a very experienced and hard-nosed wing commander, would descend on us every year – usually in the middle of the night – and Akrotiri went on to a war footing. Some of us were detached to Luqa in Malta, and the whole exercise culminated several days later with the fly-offs I described earlier. Then we waited tentatively for the assessment and verdict from the Taceval team. Nail biting and, professionally very stressful, especially for the station commander and the squadron commanders.

So, my tour in Cyprus came to an end during the summer of 1973 (just before the Turks invaded the north of the island). I returned to attend the year's course at the RAF Staff College at Bracknell driving our now beautiful Alfa Romeo 2000 Berlina saloon all the way home, with a little help from the good ship *Apollonia* from Limassol to Venice.

The Vulcans of IX and 35 Squadrons returned to the United Kingdom in 1975.

CHAPTER TWELVE

WORLDWIDE TASKS ON THE CANBERRA PR.9

JOHN BROADBENT

In August 1971, I arrived at RAF Cottesmore to begin a conversion course on to the Canberra, and where I was crewed up with Brian Carmichael. On completion of the course, we headed for Malta to join 13 Squadron.

Most of the aircrew on reconnaissance squadrons at that time, especially navigators, tended to have at least some prior operational experience on type, if not of the role. So, as first tourists, we were different. Fortunately, the squadron executive decided to keep us together as a crew rather than putting us with older hands to gain experience.

John Broadbent emerges from his navigator's position in the Canberra PR.9.

On arrival, on 24 October 1971, we needed to be converted to the Canberra PR.9 and to the reconnaissance role. The PR.9 was an extremely potent reconnaissance aircraft in its day. The usual fit of cameras was three F95s for low-level tactical reconnaissance – one forward facing, and the other two obliques mounted on opposite sides of the nose – one F96 with a 48-inch focal-length camera mounted obliquely in the centre-fuselage for

A Canberra PR.9 lands at RAF Luqa, Malta.

high-level strategic reconnaissance, and a high-resolution F49 with interchangeable six- or 12-inch focal-length lenses mounted vertically for aerial survey work. Thus, we had a lot to learn, not just about how to operate this very different mark of Canberra, but also how to exploit the capabilities of the different camera fits. So began a most enjoyable, and dare I say, successful partnership with Brian that was to last the rest of our tour, a tour that took us to many exotic locations around the world.

Almost a month of ground school followed before I was airborne, initially in a T.4, for my first three conversion missions with the squadron QFI. These were primarily aimed at area familiarisation over Malta and Sicily, and it wasn't until Exercise Four that Brian strapped in for our first taste of the PR.9. Just a few days later, after met brief, one of the flight commanders called us in and said that he needed us to take a T.4 to RAF Wyton for a reason that I can no longer recall. "Great, when?" I asked thinking that there would be a week or two to mull the problem over and get organised. "Take-off 13.00 today, don't be late," came the reply.

I can still remember the sense of panic that set in. My only experience of international flights had been a round robin from RAF Finningley via Porto to Gibraltar, the grand finale at the end of my navigator training. Those flights came after months of anticipation, weeks of planning plus, if needed, copious assistance on hand from ever helpful staff navigators. There was even one on board on the day just to ensure I didn't get lost. Now, as a supposedly fully operational navigator I was, not unreasonably, expected to take such a basic mission in my stride despite the limited planning time. The latter was reduced further because I had to dash back home to pack a bag and bid my rather bemused wife a hasty farewell. It was an abrupt introduction to the difference between the training and operational environments and I learned about squadron life from that.

By mid-December, with our conversion complete, Brian and I were tasked to take a PR.9 to Cyprus for the first of what were to be many visits to RAF Akrotiri. It was always our first refuelling point when en route to all points east, and so we became extremely familiar with its facilities. We also passed through on our return flights, and this is where the aircraft's converted bomb-bay pannier proved especially useful. It could absorb all our luggage with plenty of space to spare. This was never wasted, and over the course of a tour, many are the demi-johns of Keo brandy, a very cheap Cypriot spirit, or Kokinelli, an even cheaper Cypriot wine, that have been transferred from Akrotiri to Luqa to lubricate squadron parties.

A standard training mission of three to four hours from Luqa usually followed a lo-hi-lo profile. The initial phase would be a low-level tactical task following an overland route around Sicily and Italy. The task was to find, photograph and verbally describe up to five targets selected at random from the squadron's huge library. Once that was complete, we would climb to medium level and practise survey work. This required crews to visually navigate along three standard north-south survey lines drawn over Sicily by the squadron photographic interpreters (PIs). As the wind, especially at height, could vary significantly from day to day, every sortie posed a new problem. The objective on survey was to ensure that there were no gaps between adjacent lines, and that there was sufficient overlap between consecutive prints to ensure stereoscopic coverage when viewed by the PIs. The F49 camera didn't run continuously, instead the rate of taking images was a function of the height of the aircraft and ground speed. The crew then had to fly the lines accurately enough to maintain the lateral overlap between individual lines, and sufficient along track overlap between consecutive prints to achieve the necessary stereo coverage without wasting film or fuel. This required a high degree of map reading at height as there were no electronic devices accurate enough to help at that time. Crew cooperation was paramount. Any corrections to maintain track had to be completed in the time between camera firings. The navigator got a couple of seconds warning that the camera was about to fire, just enough time to ensure that the pilot had levelled the wings. It required great concentration, quick reactions, and very clear communication between both crew members. Once the survey element was complete – and assuming sufficient fuel remained – the mission would culminate with ship reconnaissance at low level by photographing any vessels in the waters around Malta using the F95s. All photography and the visual reports for the field targets would be assessed and scored by the squadron PIs with the results published in a league table for all to see. Personally, I loved the competitiveness that this generated and found it a great motivator to improve.

Life on the squadron simply seemed to be one novel challenge after another. I had barely been on the unit two months when Mr Mintoff, the prime minister of Malta, apparently unhappy with the rental he was receiving from HM government, invited British forces to leave the island. We received the message at squadron level whilst celebrating Christmas. Panic ensued and absolutely everything on the base had to be packed up. All on and off-base accommodation had to be vacated and contents, including private cars shipped out. The aircrew made a hurried departure with our aircraft and headed for Cyprus. We relocated to Akrotiri, where we were made to feel extremely welcome. The support staff did a great job in rehousing everyone in the few days warning they'd had of our impending arrival. I had been found an upstairs unfurnished apartment in Limassol, and I think the furniture pack we rented cost the princely sum of seven pounds a month. Cyprus in 1972 looked very different to the tourist destination that it has since become.

The squadron soon settled into its new location and the pace of operations picked up. The middle of February found us deploying to Masirah for a fortnight for what was to be the first of many such detachments during our tour. The aim was to gather military intelligence shared, I believe, with Oman. Most of the missions comprised high-level line searches off the eastern coast of Arabia using the F96 camera. The clear air at height in that part of the world, combined with the excellent resolution of the camera, meant that it was possible for the PIs to gather useful intelligence from the imagery many miles inland. The responsibility for this type of photography fell almost exclusively to the pilot as the navigator was unable to help very much given the extremely limited field of view from his

The 39 Squadron detachment at RAF Masirah, February 1972.

position in the nose. A common technique for the pilots was to draw a horizontal line with a chinagraph pencil on the inside of the canopy, then fly that line through the points of interest, in this case the coast. It might seem extremely low-tech, but through experience the pilots learned just where to position the line such that the coastline would appear at exactly the right place on the image. Very occasionally, we would be granted permission to operate over land so giving us a deeper look. Given the political sensitivity of such missions, the authority to proceed came from the very highest levels in MoD. It was challenging and rewarding work and we enjoyed those deployments to Masirah hugely.

Early May saw us taking part in the week-long Exercise Dawn Patrol, conducting low-level missions over Greece, which included refuelling stops in Crete. Barely a week later we deployed for a weekend in Nairobi. The squadron had a long-standing commitment to conduct a full aerial survey of Kenya, and this required crews, not least new arrivals such as us, to become familiar with the route and procedures involved. We passed through Masirah in both directions but, as was often the case, we needed to refuel in Tehran on our return because of unfavourable winds. This regular occurrence usually required an overnight stop. Once again, the PR.9's pannier was to prove its worth as it could hold large quantities of Kenyan fruit and coffee, and even the occasional Persian carpet.

June was to be no less busy and saw us involved in what was a never-to-be-forgotten deployment to Mashhad, a city near the border with Russia in north-east Iran. It was a fascinating place crowded with religious pilgrims there to visit the vast Holy Shrine of Imam Reza. Our purpose was to participate in Shabaz 72, an air defence exercise where our role was to act as low-level targets opposed by the Iranian air force. The local F.5 squadron made great hosts socially, which in part made up for their questionable operational effectiveness as they failed to intercept us once at low level.

There was no respite in July. Low-level missions round Italy, Sicily and Cyprus were swiftly followed by preparations for a deployment to the New Hebrides (now Vanuatu) in the South Pacific, our first formal aerial survey. We set off on 24 July and night stopped at Masirah, Gan, Tengah and on to Darwin. However, on arrival at Darwin we discovered that there was a fuel embargo in place and therefore it wouldn't be possible to refuel in Cairns our next planned stop. After a quick replan, we obtained clearance to route via Port Moresby, capital of Papua New Guinea. This created considerable curiosity and interest amongst the local population who were, understandably, not used to hosting British military aircraft.

In due course, we arrived in Port Vila, the capital of the New Hebrides, after what for me, as a first tourist, was the most fantastic and exhilarating transit packed full of new experiences. The survey took three weeks of some spectacular flying over and around the volcanic islands, several still active, that make up the territory. Our return to Malta was less exciting – it was in the back of a C-130 Hercules.

After a week in September based in Souda Bay, Crete, involved in the major NATO exercise Deep Furrow, we were informed that the British government had finally come to an agreement with Mr Mintoff about the use of the military facilities in Malta. As a result, we were warned to expect to return to Luqa early in October. The boss decided that on departure we should make a four-ship formation fly-past to show our appreciation for the outstanding support we'd received whilst based at Akrotiri. Brian and I were delighted to be chosen to participate as No. 4, wingman to Tony Wade and Pete Constant who were No. 3. Given that we flew exclusively as singletons for our day-to-day operations, this would clearly necessitate some formation practice. We completed that on 5 October and got airborne the next day, formed up to the east of Akrotiri and ran in to rehearse the fly-past down the active runway. We were then to practise some heavyweight training in the circuit.

The formation fly-past element went well, and the four aircraft broke into the circuit, and as we were approaching base turn, I noticed how quickly the speed was reducing due to our high all-up weight and called, "power!" to my pilot. At almost exactly that moment, he shouted, "I think Tony's going in!" I looked out of the tiny window I had on the port side of the aircraft just in time to see Tony's aircraft crash into the sea short of the runway leaving a huge muddy stain on the surface and not much more. There was a stunned silence in our aircraft as we tried to process what we had just seen. Moments later the airwaves were filled with calls mobilising the rescue services and so on. It is not possible to dump fuel in the PR.9, so there was no option but to remain airborne until we were down to landing weight. We decided we would fly towards Crete at low level and then turn back once confident we would be below landing weight. My logbook shows that we flew for two hours, and I don't think we spoke very much in that time as we tried to come to terms with events. We had not been on our first squadron for a year and had experienced seeing our first violent death of friends. It really was the most terrible shock but, sadly, it wasn't to be the last of my career, and it doesn't get any easier. That night we said our farewells to Tony and Pete in time-honoured fashion with the mother of all parties.

On 10 October, we departed Cyprus and flew back to Luqa. It was a tragic end to what, up until that time, had been a simply outstanding ten months and a terrible conclusion to my first year on the squadron.

January 1973 was no less hectic and, on the 20th, we deployed to Sanaa, capital of Yemen, for Operation Hayrack, the second major aerial survey of my tour to date. We took three aircraft, an appropriate number of spare aircrews, plus a full contingent of engineers. It had proved impossible to secure accommodation in the city deemed of a suitable standard, so we set up a tented encampment about ten miles away on the high jebel. The airfield itself was in the process of being modernised by a German contractor and at the time still lacked all but the most basic amenities. Consequently, a very substantial number of C-130 flights were needed to bring in everything we needed, including air traffic controllers, fire engines and, of course, copious quantities of the amber nectar. The army provided a full field kitchen, and they really knew how to make the most of compo rations, albeit supplemented by local fresh produce.

The hot rumour on the squadron at the time was that the UK had secured the task in the face of competition from the Soviet Union. The latter had allegedly offered to produce the updated paper mapping in a couple of years – at that time, a not unreasonable timescale. The UK on the other had offered maps in less than twelve months. Whether that was true or not, it was the case that, on each survey mission, crews carried two sets of maps. One was based on decades-old data and was the one we were supposed to be updating. It was to be shown to any Yemeni official who asked to see what we were using in the air. The other set, kept concealed until airborne, was the one we used to conduct the survey. It was much more recent, and probably based on satellite imagery. It would have been significantly quicker to update that set than the older one. The squadron junta felt there had to be a reason for all the subterfuge and, in the absence of anything official, the rumour mill filled the void.

It is not uncommon on major detachments such as this, that the boss would arrange for a cocktail party to entertain and show our appreciation of embassy staff, as well as local dignitaries and expats. The venue was our large dining tent and, in deference to local sensitivities, the boss decreed that it should be soft drinks only – at least until all our Arab guests had departed. We were all allocated guests to host and one of mine was the Yemeni destined to become the manager of the new international airport. After half an hour or so of drinking orange juice, he asked if we had anything stronger. I was non-committal but took his request to the boss, to be told it wasn't the first he'd received in a similar vein. He in turn checked with the ambassador, and between them they sanctioned the opening of the bar. A fun evening ensued, but it did result in the ejection of one of the corporal chefs after he attempted to provide some unscheduled entertainment. Having first divested himself of most of his clothing, and to his own rendering of 'Hold 'em Down You Zulu Warrior' … I will leave the reader to imagine what followed.

It was possible, when not on the flying programme, to visit Sanaa, an ancient, traditional Arab city that I found absolutely fascinating. The advice from the embassy was that it was usually perfectly safe although it was best to avoid Fridays, especially around midday prayer time. This was when local feelings were at their most delicate, and that public amputations, although not at all common, were most likely to be performed. Secondly, if using a local taxi, it was wisest to travel as early as possible in the day. Apparently, a significant proportion of the population, which included taxi drivers, was addicted to chewing large quantities of Qat, a local shrub with dark green leaves not unlike privet in appearance. It had stimulant properties, which in due course rendered the drivers rather too carefree for comfort. The evidence was often only too apparent: heavy users sported green teeth and permanently stretched cheeks from the constant chewing.

It took us just three weeks to obtain full photographic coverage before we were packing up and heading homebound. One unusual aspect of the recovery on this occasion was our routing with a refuelling stop at Cairo, the first time British military aircraft had been welcomed since 1956. The boss went to considerable lengths to stress the importance of ensuring that none of the aircraft should be carrying anything that might be deemed controversial. He was especially keen that none of our camera magazines should have film loaded, as we were bound to be subjected to thorough checks. We were, however, encouraged to make a note of any military equipment that might be visible around the base. Two hours on the ground and we were on our way again.

At the end of June, we departed for Hong Kong and Operation Chaucer. Like the Kenyan survey, this was a long-standing regular commitment to survey with the F49 camera the vibrant, ever-changing city and the New Territories whilst, at the same time, taking a surreptitious look over the border into mainland China using the F96. We followed our now-familiar route to Masirah, and then continued to night stop in Gan and Singapore. Bearing in mind the Vietnam War was still in progress, I suppose that we shouldn't have been that surprised to find that our flight plan for the last leg to Hong Kong was initially refused by ATC. When we enquired why, we were told that military traffic en route was not allowed below 40,000 feet. Well, that didn't trouble a PR.9, so I re-submitted for Flight Level 410, and we were cleared to depart. It was very interesting to listen to the constant chatter on guard frequency from the American crews engaged on offensive missions over Vietnam.

Unfortunately, that wasn't to be the end of our troubles. On 9 July, we were airborne for our third survey mission mindful that the met report for the day warned of the approach of a typhoon later in the afternoon. We climbed to survey height and had just started the task when ATC informed us that the storm was making more rapid progress than anticipated and that we should consider recovering as soon as possible. Once down to landing weight, we began our approach in increasingly turbulent conditions that caused an uncomfortable amount of airframe buffeting. The approach to runway 13 at Kai Tak could only be executed in visual flight conditions and was notoriously challenging even in the most benign conditions. Not only was there high ground to the south and west, but it also required a sharp right turn at 650 feet over the city at the 'checkerboard' to line up on the runway centreline. To make matters worse, the electrical nature of the approaching storm was playing havoc with my rudimentary nav aids making steering to reporting points unreliable. With considerable skill and no little luck, we landed safely, and it was a very relieved junior crew who made their way to sign in. We subsequently discovered that one of the structures securing the rudder had failed, possibly because of the battering it had received during the approach. It turned out that this was such an unusual failure that there were no spares and so we would be grounded pending manufacture of a new part. In the end it took two weeks, during which time we were let loose on an unsuspecting Hong Kong. The aircraft became serviceable again on 24 July by which time we were due to return to Malta. We had no option but to reverse our outbound route, add in a night stop in Tehran and head back to Luqa. I have very happy memories of that fortnight in Hong Kong.

The rest of 1973 was filled with the usual bread and butter training sorties over Sicily, Italy, and Greece. These were interspersed between four detachments to Akrotiri for a variety of exercises, plus two to Wyton, one of which was a week for night photography over UK overland ranges, and another flying visit to Mashhad in Iran. Much later I was to discover an unexpected intelligence output from the routine ship reconnaissance that the squadron had conducted around Malta in the run-up to the Arab-Israeli war that took place in October. Post-war analysis of our imagery indicated that during that period there had been a measurable increase in commercial shipping destined for Egypt. Some of the deck cargo was subsequently identified from its distinctive signature as containing Russian-built military equipment. This was just one of many intelligence indicators that, taken together, suggested that something was afoot. Unfortunately, their true significance wasn't recognised in time to predict the pre-emptive strike.

1974 began with a night stop in Gibraltar for more ship reconnaissance, followed in March by another ten-day detachment to Masirah. It was also about this time that the squadron began to re-equip with the PR.7. I thought then that this was a massive retrograde step in capability, but was probably inevitable given the increasingly poor availability of our PR.9s. All new aircrew arrivals at that time went straight to the PR.7 so I was relieved that Brian and I were one of the crews that remained assigned to the PR.9. Although I did subsequently acquire almost twenty hours on the PR.7, including a detachment to the UK for some low-level reconnaissance where I enjoyed being able to see out for once, overall I was not a fan.

As spring turned into summer, we took part in several major NATO exercises and then, in July we were awarded our command crew category by the C-in-C. The boss was at pains to tell us not to get an overly inflated sense of our own importance. He considered it had been awarded, at least in part, as forgiveness from the C-in-C for the disrespectful way he felt that the squadron had treated him the previous year when he was guest of honour at a dining-in night at Luqa. I remember the night very well. In the week leading up to the dining night, we had been one of several crews detached to Cyprus for an air-defence exercise. It finished on a Friday afternoon, and we had all hurriedly flown back to Malta, so it was something of a panic to get ready for the dining-in night. I think we were vaguely aware that there would be a VIP guest, but I for one hadn't really registered just how important that guest might be. After a mad dash home to get changed and a few brandy sours before taking our seats for the dinner, we were in high spirits. After more wine and port, spirits were even greater. One of the squadron navigators, who shall remain nameless, had what the junior junta all agreed was a highly amusing party trick. He would slide out of his chair during the meal, crawl beneath the tables avoiding diners' feet, and then tie together as many of the shoelaces of the senior officers on the top table as he could before reversing direction. Depending on how long this took, and if it could be done surreptitiously, his immediate neighbours thought it a jolly jape to spirit his chair away and close the resulting gap. If successful, he would then have no option but to lie under the table until the dining room was vacated. Childish perhaps, but irresistibly entertaining assuming one had consumed enough alcohol. On this occasion, the effectiveness of the shoelace tying only became apparent when the C-in-C stood up to make his speech at which point he found that his shoes had mysteriously become connected together. To say he had a sense of humour failure is to significantly understate his reaction. The net result was that the boss was required to call the entire officer contingent of the squadron to attend a mass lecture at dawn the following morning. I can only admire the boss's masterful delivery of the reprimand and we left suitably chastened.

Be that as it may, because of our new status, Brian and I were almost immediately involved in a particularly unusual task. We were called to Akrotiri and briefed that the *Moskva* helicopter carrier (not to be confused with the cruiser of the same name sunk by the Ukrainians in the Black Sea in 2022) had been spotted whilst passing through the Bosphorus into the Mediterranean. What was unusual about this was not the ship but what appeared to be preparation of her flight deck like that undertaken by the RN for Sea Harrier operations. The concern was that perhaps the ship was about to embark the Yak 38 'Forger' – the Soviet equivalent to the Harrier. To gather further intelligence, we were dispatched on 26 July to take as close a look at low level as was legally allowed. The visibility that day was unseasonably poor, just a few kilometres in mist below a low cloud base. As the PR.9 had no radar, to ensure that we would find the *Moskva*, a Nimrod operating at medium level provided us with regular steers. In due course the *Moskva* appeared out of the peasouper, and we sped past close enough to collect some excellent imagery with the F95. I did have one moment of disquiet seeing all the ship's defensive armament tracking us. Clearly the ship had been aware of our approach despite the weather. If we had had a radar-warning receiver in 1974, I am guessing that it would have been going frantic long before we appeared over the visual horizon.

Broadbent Could Get Worse!

THE Squadron was visited last month by the AOC-in-C, Air Marshal Sir John Aitken, who presented Command Category Awards to Flt Lt Brian Carmichael and Fg Off John Broadbent. This award was particularly well-deserved and all the Squadron congratulate them. Needless to say they are both modest gentlemen who would not let such an award go to their heads — but John was heard the next day to reply rather haughtily to a request to do a compass swing with; "I'm Awfullly sorry, I'm afraid Command navigators don't do compass swings'. It's going to be even worse when he becomes a Flight Lieutenant.

The RAF *Luqa* magazine 'reports' John Broadbent's award of his command category.

All too soon the end of my tour loomed, but on 2 September we were fortunate enough to be allocated a last 'Island Ranger' to Nairobi. The route outbound was very unusual as we were given clearance to refuel in Cairo. This enabled us to fly on directly to Nairobi and thus reduce the transit time considerably, which in turn gave us more time in Kenya. We remained in Nairobi for a week and then, unable to repeat the outbound route, came home via Djibouti instead. It was a wonderful way to finish what was simply an outstanding tour in which I accrued 1,115 flying hours. I operated in places from Gibraltar in the west to the New Hebrides in the east. It included locations few would have the chance to visit even today – Iran, North Yemen, Papua New Guinea, and Djibouti to mention but four. It also took in places that are now exotic tourist destinations such as Egypt, the UAE, Oman, Kenya, Vanuatu, Australia, the Maldives, Singapore, Hong Kong and Australia. I should not forget to mention ten months detached to Cyprus and extensive land aways throughout the Mediterranean. Simply Awesome!

POSTSCRIPT

I had my final sortie on 13 Squadron later in September and not long after started my conversion to the Buccaneer S.2B in the strike/attack role. That was followed in 1982 by conversion to the Tornado GR1/GR1A. I had a terrific time on all three aircraft and have fond memories of all of them but none better nor more vivid than those I have of that tour on the PR.9 in the early 1970s. Having said that, I confess that the intervening years when I was operating aircraft where it was possible to admire the view whilst being whisked around at low level by extremely competent pilots rather dampened my enthusiasm for 'the Black Hole of Calcutta', which is how the navigator's position in the enclosed nose of the PR.9 was described by some. On the other hand, it was specifically designed for the role, all the navigation and camera controls were right to hand and thus it made for an excellent professional environment.

I had the great good fortune to be associated with the Canberra PR.9 at both ends of my RAF career. It was my first operational aircraft, and it was also my last because my final tour in the RAF was as Station Commander RAF Marham, at that time the RAF's 'Centre of Excellence for Reconnaissance'. It was home to two Tornado GR1A squadrons but also No. 39 (1PRU), operating the last of the RAF's Canberra fleet of T.4s and PR.9s – and still doing extremely useful operational work.

Looking back through my logbook at those years now, I realise that I was hugely fortunate to be on the squadron as an era was ending. Those were halcyon days when, in addition to the UK home commands, we still had separate Near and Far East Air Forces and an RAF Germany. Nor was the hand of supervision nearly so heavy as it was to become by the time I joined the Tornado Force a few years later. A first tourist crew such as Brian and I were regularly dispatched to exotic locations such as Kenya and Hong Kong. Whilst there, we would be largely unsighted by squadron hierarchy at base beyond a daily sitrep. and reliant completely on our own self-discipline and professionalism. Therein lies a story or two as well, but the fact is that we came to no great harm, we acquired massive amounts of airmanship in a very short space of time, and it was character-building stuff. It was certainly to stand me in good stead later in my career – and how I enjoyed those carefree days.

CHAPTER THIRTEEN

ADEN HUNTERS

PETER STURT

I WAS THE YOUNGEST OF FOUR children with a father who survived World War One flying DH4s as a captain in the then RFC, and subsequently RAF. Two of my siblings preceded me into the RAF. I joined the RAF in 1959 at the RAF College Cranwell, and three years later, as a newly fledged pilot officer with my wings qualified on the Jet Provost. I did advanced flying for six months on the Vampire at RAF Valley and then operational flying conversion to the Hunter F.6 at RAF Chivenor. Six months later, in September 1963, I was posted to 8 Squadron at RAF Khormaksar in Aden, flying the Hunter FGA.9.

Peter Sturt prepares to board a Buccaneer which he flew after leaving Aden.

SHORT HISTORY OF ADEN

In 1839 Britain captured the town of Aden (now part of Yemen) in the south of the Arabian Peninsula. The occupation of Aden was a strategic rather than commercial undertaking, guarding the lines of communication with India. With British Somaliland on the 'Horn of Africa', Aden provided control of the entrance to the Red Sea. Following the opening of the Suez Canal in 1869, Britain established protectorates in the hinterland of South Arabia to act as a buffer against the Ottomans who occupied Yemen. In 1937 Aden became a crown colony.

Britain granted independence in February 1959 to the FRA (Federation of Arab Amirates of the South), which was formed from the Aden colony and the surrounding protectorates, to stabilise the region, which had been dogged by years of unrest fuelled by Arab nationalism and anti-colonialism.

UNIFIED COMMAND

The Middle East Air Force was formed on 1 March 1961 as a unified command from HQ British Forces Arabian Peninsula having been originally named Middle East Command (Aden), the RAF element being named Air Forces Middle East. With the impending British withdrawal from Aden, authority for the command moved to Muharraq (Bahrain) on 1 September 1967, and it was renamed HQ British Forces Gulf.

The Middle East Air Force territory covered a vast swathe of Arabia, from Bahrain Island in the Persian Gulf in the north, down through the Trucial States, Muscat and Oman, the South Arabian Federation and across the Horn of Africa to Kenya, a distance of some 2,500 miles. The centre of operations was located at RAF Khormaksar in the Aden Protectorate. Units based at RAF Muharraq on Bahrain Island in the northern sector maintained constant vigilance as to the behaviour and intentions of nearby Gulf states, Iraq in particular. Intermediate replenishment airfields (known as route stations) were provided at Sharjah, Masirah Island, Salalah and Riyan. To the south lay Ethiopia and Somalia and beyond that, the lush green plains of Kenya where, close to the city of Nairobi, RAF Eastleigh provided the main operating base for the East Africa region.

In 1962 the British government announced that a permanent British garrison would be maintained in Aden. Yet by 1967, the British presence was untenable and was forced to withdraw from the colony.

THE HUNTER FGA.9

The Hunter was the standard single-seat fighter in the RAF Middle East Command with 8, 43 and 208 Squadrons based at Khormaksar. All three squadrons, some 40 aircraft and a small flight of Hunter FR.10s (reconnaissance variant), came under the control of Tactical Wing. A major reorganisation of the Hunter Wing was introduced with the reformation of 1417 Flight in May 1963. This 'new' unit took charge of all four of 8 Squadron's FR.10s, its two T.7s and a T.7 each from 43 and 208 Squadrons. The squadron markings were removed from the FR.10s and the initials of the four most senior pilots applied to the tailfins and nose doors in light blue. The whole Hunter Wing, commanded by a wing commander, was responsible to the station commander for the coordination and control of tactical flying at Khormaksar. A new centralised servicing system was introduced during the month of June

A Hunter FGA.9 over the Aden hinterland.

1964. All second-line Hunter ground crews were pooled into a dedicated team known as Rectification Flight and were tasked to perform specific servicing tasks on all Hunter Wing aircraft. This system was not popular with the squadron commanders and some of the engineers since it took away their hard-won squadron identities. As a result, it was quietly forgotten after a few months.

The aircraft had a ceiling of 50,000 ft, a range with drop tanks of 1,650 nm, and a top speed of around 600 knots at sea level. Our routine flying would be at low level 200 ft AGL and 420 knots. Where necessary, to reach more distant areas we would do hi-lo-hi sorties since the Hunter, and all other jet engines, were more fuel efficient at high levels.

The FGA.9 was equipped with four 30-mm Aden cannons for the air-to-ground and air-to-air role and World War Two-vintage HE 60-lb rockets for the air-to-ground role. The latter were fired from rails attached to the outer pylons and occasionally the inboard ones as well. In this event our flight time was limited since we lost the benefit of extra fuel in the drop tanks on the inboard and outboard pylons. We had an air-to-ground firing range close to Khormaksar a few miles to the north-east of the airfield and practised our skills there on a regular basis.

Low-level, cross-country interceptions, operational rocket-projectile, air-to-ground firing and high-level battle formations and tactics were practised on a routine basis. Sorties would also encompass elements and variations of the following: aerobatics, instrument flying practice, practice emergencies and diversions (our nearest and practically only one was at Djibouti), and general handling at high, medium, and low altitudes over the sea and overland. The air-to-air threat, albeit a low one, was from Egyptian MiG-15s operating from the Yemen and Yemeni Yak II ground-attack aircraft.

WORK UP TRAINING

At long last after some four years of training I found myself embedded into a fighter ground-attack squadron and flying an exciting, exhilarating and highly manoeuvrable aircraft. All this was possible in an area largely devoid of any restrictions in our areas of operations. The squadron policy was to get new pilots up to operational standards as quickly as possible. This meant for me that I was given flying priority over other pilots on the squadron. As a result, after two months I had achieved sixty hours of flying and was declared operational.

One of my work-up sorties was as No. 2 to my flight commander that involved a 250-ft low-level area familiarisation sortie. The brief was to stick close for tactical manoeuvres at low level and to keep out of his slip stream. All was going well until on one very hard turn to starboard, with near ninety degrees of bank, I found difficulty in seeing him, so I dropped slightly low – in doing so I hit his slipstream at high g around 420 knots and immediately was flipped on my back. I was now staring at the very close hard, rocky, and sandy desert floor. Full left aileron was having no effect, so I completed the roll to the right, pulled hard away from the ground and saw my leader some 3–400 yards away to my right. "What happened to you?" When I explained, he commented, "Well you will not do that again will you". I certainly frightened myself on that occasion and learnt a very hard and near fatal lesson.

Flying at low level was exhilarating – especially wadi bashing which we all did since it had an operational role. There was one particularly steep wadi with vertical cliffs and a right-angle bend which we all had to negotiate in our work-up sorties. We flew this initially at the top walls of the wadi and then followed our leader at a safe distance (some 4–500 yards) in doing so demonstrating how to negotiate the right-hand bend safely at the bottom of the wadi whilst pulling 4–5 g at 420 knots. With ninety degrees of bank, it was not possible to see the far wall of the wadi. We just had to be confident of the capabilities of our aircraft and the fact that the aircraft in front had just done it.

DETACHMENTS

Squadron detachments to airfields along the south coast to RAF Masirah, past Oman, RAF Sharjah and RAF Muharraq in Bahrain in the early 1960s were commonplace. For a few years, each Hunter squadron detached en mass to reinforce Bahrain for two months, being replaced by a sister squadron in constant rotation.

This was a welcome break from the sand, dust, and heat of Aden, but at times the humidity became unbearable approaching 100 per cent. There was however a very enjoyable swimming pool close to the officers' mess. One pilot decided to liven things up a bit by bringing a donkey into the mess. One winter over Christmas there was ice on the palm trees. This surprised us all but apparently the Shamal was blowing cold air from the mountains in Persia.

To reach Bahrain we had to refuel at RAF Masirah, and sometimes also at RAF Sharjah (now totally dwarfed by Dubai) and occasionally at RAF Salalah en route to RAF Masirah.

OPERATIONS

There were numerous operations flown by all three squadrons in the Aden area, especially in the period 1963–1965 when incursions by Yemeni aircraft and insurgent ground forces were starting to increase. Below is a snapshot of some of these operations.

Following a border incident in the Beihan area early in 1963, 8 Squadron brought two FGA.9s and two FR.10s to one-hour's readiness with the rest of the unit's aircraft on two hours. Even so, a Yemeni Yak II aircraft was able to cross the border and strafe the Ayn airstrip and return unmolested by the patrolling pair who were at 7,000 feet. Four days later, the tactics were revised so that one of the pair flew at 7,000 ft and the second at 2,000 ft. This became a routine for the Hunter squadrons for many months when border incidents in the Beihan area were expected. Operation Ranjii kept two Hunters on standby to react to hostile aircraft incursions over the Yemeni border. These were also utilised to provide top-cover patrols to the Beihan area when resupply to units on the ground at the Ayn airstrip by the Beverley squadron were in progress.

Continuous dawn-till-dusk top-cover sorties were flown over the Beihan area and in support of Operation Jackknife in the Wadi Ayn area for the first week of February 1963. With the re-location of 43 Squadron from Cyprus to Khormaksar early in the month, the pressure on 8 and 208 Squadrons' aircraft and crews was greatly eased. The short-term deployment of Sea Vixens from HMS *Centaur* on the border patrols, further helped the squadrons recover.

Although the FR.10s were ideal for photographing prospective targets and the results of subsequent attacks, Canberra PR.7s and PR.9s were sent to Aden on short detachments from the UK, Germany, Malta and Cyprus to photograph the whole area. Two Hunters were usually assigned as armed escorts to the Canberras on these sorties.

A flare-up in early July between dissidents at Hilyan, during which shots were exchanged, triggered the dispatch of Hunters on leaflet drops, backed up by flag-wave sorties the next day. As this made no impression, a force of Hunters from both 8 and 208 Squadrons carried out strikes on various targets around the local fort on the 9th and 10th. An old feud between the Rabizi and Illahi tribes in the Upper Aulaqi area was threatening to flare up once again and several flag-waves were flown on 27 and 28 July to try to cool things down. More leaflet drops were carried out on surrounding tribes on the 29th to warn them not to get involved and

A Hunter 'upcountry' in the Aden Protectorate.

this seemed to have the desired effect; the area quietened down. A reported increase in gun-running activities by various upcountry tribes, saw the introduction of regular reconnaissance flights to seek them out.

Visits by Royal Navy ships to Aden were often used to extend Hunter pilot training and several rocket and gun-firing sorties were flown against splash targets towed by HMS *Victorious*. Trials were carried out between 6 and 13 September with the recently shipborne-installed UHF radio equipment, in which the aircraft were flown at high speed and at low level to overcook their black boxes, but a letter of congratulations proved that we 'failed'.

The new year began with operations in support of FRA units which were heavily engaged in Operation Nutcracker against dissident tribesmen in the Jebel Radfan area. No. 8 Squadron flew forty top-cover sorties and twenty-six attacks on ground targets in the first week of January. February continued in much the same vein with thirty-one operational sorties flown on Operation Rustum, seven as top cover and twenty-four on attacks against buildings. Seven further sorties were flown in support of an FRA company being fired on from a village east of Dhala on the 24th, a leaflet drop being followed by a four-ship cannon strike and a pair using rocket projectiles.

The attack on Fort Harib on 28 March 1964 was arguably the highest profile incident involving the Hunter during its eight years of operational service in the Middle East. Three Khormaksar-based units, 8 and 43 Squadrons and 1417 Flight, carried out the attack and more details can be found in the respective squadron operations record books (ORBs). The operation caused a political storm back in the UK and in the United Nations.

Sorties were flown in support of army operations in the Radfan which had, as their major objective, the clearing out of rebels from the area. Aircraft were often scrambled to help ground troops out of difficult situations, and several rocket and cannon attacks were made on houses and crops to flush out troublemakers. Sorties continued to be flown in the year to intercept unknown aircraft heading for the border, but the target(s) turned back without crossing over. Disorder amongst tribesmen in the Wadi Jirdan and Wadi Habban was quelled by eight flag-wave sorties over two days in March. Later, two Hunters were scrambled to support FRA units, attacking dissidents on the Dhala road under forward air control (FAC).

Relentless demands for ground support in the Radfan area, and Beihan patrols continued through July until the end of the year. In addition to attacks by Hunters, Shackletons of 37 Squadron began to play a greater role with the Hunters marking targets for Shackleton bomb-aimers.

The tempo of the preceding two years continued with operational sorties flown comprising: Beihan patrols, flag waves, recces, leaflet drops, armed recces near Dhala and FAC-controlled strikes in support of ground troops moving up the Dhala road. However, the internal security situation in Aden remained poor and on 29 May 1965 an Aden Airways DC-3 aircraft was destroyed on the ground at Khormaksar Airport. As this was almost certainly an act of sabotage, extra vigilance was called for especially with pre-flight inspections on all RAF aircraft on the station.

On every occasion during Ramadan our squadron coffee boy always disappeared to his family up country in the Radfan where he claimed to live. After Radfan, he would always reappear as usual and on one occasion he was very excited to say that he was watching us attacking his uncle's house. He was very sad that we had not hurt or killed his uncle because he did not like him at all. Clearly, he gained in stature with his friends in working for those who flew the mighty Hunter.

42 SQUADRON AFFILIATION 10 MARCH 1965

This was an interesting sortie that included formating on a Shackleton. The OC 42 Squadron was flying the Shackleton while our flight of four Hunters from 8 Squadron practised some close formation on his aircraft. I was flying XF552-D. The main thing I remember from that sortie was when we formed up in echelon port on the port wing of the Shackleton prior to landing at Masirah. The intention was for the fifth Hunter to take a pretty photo of the four peeling neatly off to the left. Unfortunately, OC 42 decided to call the break rather than our formation leader and he did it by saying 'standby to break', 'break 1', 'break 2', 'break 3', etc., in that order. Our whole formation twitched horribly. I was No. 3 and when I saw the two aircraft on my right banked towards me, I heaved up and away out of it. It took some time to get us back into any semblance of order and we did not see OC 42 after that. I think he took his aircraft on to Sharjah or Bahrain.

September dawned with more of a flourish with a fire-power demonstration at Jebel Hanak at midday to deter potential collaborators. On the early afternoon of 1 September, units of Nos. 4 and 5 FRA had met strong opposition in the Jebel Khuder area, the latter group becoming pinned downed by dissidents. To relieve the situation, a five-ship strike was mounted on the dissidents. Air support was provided to give top cover as a further series of eight strikes on the same target was mounted through the early evening hours.

In what was a joint operation, two 37 Squadron Shackletons carried out fire-power demonstrations, each aircraft dropping fifteen 1,000-lb bombs on Jebel Hanak, close to Ataq, where twenty-two houses had been fired upon by dissidents under the leadership of Ahmed Bubakr Bin Saridim, a well-known troublemaker and a recent arrival in South Arabia from the Yemen. It was hoped that the demonstration would persuade local inhabitants not to support the dissidents who would, without support, have soon faded away.

Five Hunters of 8 Squadron escort a Shackleton MR.3 of 42 Squadron over Aden.

Combined RN/RAF fly-past over Aden in May 1967.

Having had little effect, the action continued at first light the next day from where it left off on the previous day, with three pairs mounting strikes on Jebel Khuder over a ninety-minute period to cover the withdrawal and evacuation of FRA casualties by helicopter. As a result of the air action, five dissidents, including a prominent leader, were known to have been killed and three wounded. The next day fifteen strike sorties were flown to attack and destroy two houses near Wilan that belonged to a well-known collaborator in the Wadi Maraban. Further flag-waving, armed-recce and strike sorties continued until the end of the year.

After two years of exhilarating flying an outstanding aircraft, and in a demanding environment, my time in the sweltering heat of Aden came to an end early in September 1965. I headed off to Tern Hill for two months on the JC&SS (Junior Command and Staff Studies) course and then on to Chivenor as a tactical flying instructor on the staff of 234 Squadron, little realising that I would be flying off Aden again within two years of leaving, but in a very different situation.

ADEN FINALE

In 1967 the decision was made to withdraw all British forces from Aden and South Arabia when the country was to achieve independence in January 1968. In May 1967 I was flying off Aden again, this time in a Buccaneer of 809 Squadron having volunteered for an exchange tour appointment with the Fleet Air Arm. The two aircraft carriers, HMS *Hermes* and *Victorious* were en route to and from the Far East respectively. The opportunity was taken to indulge in cooperation exercises with the Khormaksar-based Hunter squadrons. As a show of strength, several close formations and fire-power demonstrations were flown with them, over Aden and the Protectorate including a massive fly-past of fifty-two aircraft from the Sea Vixens and Buccaneers plus the Khormaksar wing of Hunters. Yours truly was in this formation.

This took place on 17 May 1967. The formation was led by Sea Vixens from 892 and 893 Squadrons, followed by Buccaneers from 801 and 809 Squadrons with 8, 43, 208 Squadrons and 1417 Flight Hunters bringing up the rear. This was quite a memorable flight since the Vixen-lead pilot was not that familiar with leading large formations. During the final run in to Aden he decided that he was a few seconds too early and decided to slow down from the run-in speed of 350 knots. Unfortunately, he chopped his throttle to slow down rapidly to about 250 kts and the concertina affect as this rippled down to the back end of the formation (especially the Hunters at the very back) was again quite interesting. Fortunately, the whole formation achieved some order by the time it overflew Aden. I suspect the comments in the Jungle bar that evening amongst the Hunter fraternity were somewhat blue.

So, I finally bade farewell to Aden.

CHAPTER FOURTEEN

KEEPING THEM GOING

RAY DEACON

HAVING PASSED OUT FROM COSFORD as a Boy Entrant 'Brat' in December 1959, followed by two-and-a-half years working on Vampire T.11s at the Central Flying School, RAF Little Rissington, I was posted to the dreaded Aden. Most postings to the Middle East began in the comfort of a BUA Britannia but on 10 April 1962, I joined a group of thirty airmen for a ten-day cruise aboard HMT *Nevasa*. At least the tally was deducted from our 730-day tour duration. On arrival in Aden, I was assigned to 8 Squadron and took up residence in the aptly named Hunter block, one of three recently constructed with four-man, air-conditioned rooms.

Respect for the way in which airmen went about their task of keeping a regular supply of serviceable aircraft, has been acknowledged by pilots over the years. As the Hunter was intended for service in the relative cool, clear-air climate in Europe, maintaining availability in the hot, humid, and dusty climate of the Middle East presented new challenges for squadron ground crews.

In 1962, 8 and 208 Squadrons were the Hunter units based at RAF Khormaksar. No. 8 Squadron was equipped with fourteen FGA.9s for air defence and ground attack, four FR.10s for fighter/photo-reconnaissance and a pair of T.7 trainers, with a further twelve

Ray Deacon ready for his flight in a Hunter T.7.

An airman marshals in an 8 Squadron aircraft at Khormaksar as a second airman prepares to attach the access steps to the cockpit.

FGA.9s and a T.7 allocated to 208 Squadron. The squadrons provided support for a large, diverse force of British and Arab army units, spread throughout the hostile, mountainous hinterland in Aden and the whole of the Middle East Command (MEC) area.

GULF CRISIS

Tranquillity was a rare experience in Arabia. In June 1961, General Kassim of Iraq aligned his forces along the border with Kuwait in preparation for an invasion into the tiny state with its lucrative oil fields. In response, the British government ordered 8 Squadron to fly up to Bahrain, 250 miles east of Kuwait, and prepare to defend the state in the event of an Iraqi attack. No. 208 Squadron, which at that time was based in Kenya, was ordered up to Bahrain to reinforce the British air defence and strike capability, as a dozen Canberras were dispatched from Germany to RAF Sharjah. Troop reinforcements were flown into the area from Europe and other MEC bases, as a third Hunter squadron, No. 43, moved from Leuchars to Cyprus. The errant general was left in no doubt that Britain would defend Kuwait and, although he backed down, it was decided that the Persian Gulf would never again be left undefended.

OPERATION LONGSTOP

Concerned that a total withdrawal of its forces from the Gulf could tempt Iraq to make another attempt, the British government wished to maintain a squadron of Hunters and a pair of Canberra PR.9s at Bahrain. However, the Sheikh of Bahrain refused permission for the permanent basing of offensive aircraft on his island but agreed to a programme of rotating the Hunter squadrons from Khormaksar. This was implemented under the name Operation Longstop. From November 1961, 8 and 208 Squadron aircraft and personnel were flown up to Bahrain every other month on a two-month detachment and that was the routine when I arrived. Ground crews and equipment were transported by Beverley and later, by Argosy. The climate in the Gulf differed notably to that in Aden, hotter and more humid in the summer yet cool enough to warrant a change into working blue (uniform) in winter. During the Gulf detachment, the squadrons spent two weeks on APC at Sharjah to make use of the range at Jeb-a-Jib.

LIFE ON THE LINE

Due to the intense heat, the normal working day at Khormaksar was from 0700 to 1300, followed by lunch and a two-hour siesta. This would quickly change when air support was requested by army units operating upcountry. Assigned to the Hunter line as one of a handful of wireless mechanics, my job was to ensure that the radio equipment on our Hunters was working properly, before and after each sortie. Inspecting aerials on top of the fuselage could be a challenge, requiring mastery of a technique if blistering of the knees from hot, metal drop tanks was to be avoided. The knack was to run towards the wing, bounce a knee on the drop tank and haul oneself up and onto the wing using the handily placed saw-tooth leading edge as a handgrip. Different styles were employed, and

Khormaksar's Strike Wing of 8, 208 and 37 Squadrons.

a mistimed attempt made good spectator sport as the unfortunate ended up in a painful heap on the pan. During a Bahraini summer, a Hunter wing was hot enough to fry an egg; one intrepid pilot taking twenty minutes to prove the point.

One of the most difficult jobs for radio mechanics was working inside the cramped Hunter radio bay when temperatures could reach 140°F. The bay was so confined that it was impossible to turn or twist. Changing one of the two 30-lb VHF sets was not too much of a problem as they could be rested on a shoulder as you lowered yourself down and out of the bay, but it was a different story once these 'lightweights' were superseded by a single, 56-lb Collins ARC52 UHF unit. Changing one of them could be a struggle, the sweat literally pouring down one's legs, but it did contribute to our overall fitness.

The radio/radar guys had it somewhat easier than some trades; the armourers, engines, and airframes guys particularly, their jobs demanding much longer in physical contact with the airframe. The energy and enthusiasm of the armourers deserves a special mention, especially when changing gun packs and loading rockets during operations and sessions on the range. Working in teams, each would try to beat the 'record' for changing a gun pack which, if I recall correctly, was a little over eight minutes. I offered to help once. Lowering an empty pack onto the trolley was straightforward but winching a fully armed pack back up to the locking points was a different ball game. Nearing the top, with arms aching, I ran out of steam, forcing the other two to await my recovery. Something I had to make up for in the Camel Club bar that evening.

Armourers attach a 60-lb HE three-inch rocket to a Hunter.

SAFETY FIRST

From the first day of training, the significance of flight safety was drummed into everyone. On front-line squadrons, where aircraft were regularly armed, additional safeguards were employed. The start-up procedure on the Hunter FGA.9/FR.10 was straightforward but an alert mind and an awareness of the movement of others around you were essential. One airman was assigned to each Hunter, and it was his responsibility to ensure all ground equipment was clear of the aircraft and that a ladder was attached to the aircraft while the pilot carried out his pre-flight checks. Once he was seated in the cockpit, the parachute and ejector seat straps were aligned over his shoulders and, on request, his bone dome handed to him. The radio connector was then plugged into its socket and the ejector seat safety pin removed, shown to the pilot, and stowed in a pocket in the side of the seat. Having climbed down, the ladder was removed and positioned in a safe place on the pan as the pilot carried out his pre-start checks.

When ready, the pilot gave a twirl of a forefinger and, having checked that the area was still clear behind, the airman reciprocated with his own forefinger gesture. Pressing the start button initiated the automated engine-start sequence. The highly volatile fuel known as Avpin was used to start the Rolls-Royce Avon 207-series engine on these Hunters, which emitted a cloud of acrid gas on ignition accompanied by a flame from an exhaust duct underneath the fuselage. As this had caused several fires in the UK, the starter-bay door was left open until the starting cycle completed. An asbestos glove was used by airmen to smother any flames. The door was then closed, and the airman stood ready to remove the nosewheel chock and marshal the aircraft out.

PHANTOM ON THE TRIGGER!

Aden folklore told of an incident during the early period of Hunter operations in Aden, that was to cause the introduction of a new safety measure. It was said that the cannon of a cannon-armed Hunter parked on the pan in the intense heat, fired off several rounds of ammunition into the wall of an army barracks on the far side of the Khormaksar runway. It was fortunate that no aircraft were taking off or landing at the time and no casualties were reported from the barracks. The reality, however, was that no one had been working on or near the aircraft at the time and the incident could have resulted in serious consequences. On investigation, it was found that excessive heat had induced a short in the gun-firing circuit. A microswitch in the port undercarriage bay was intended to break a circuit to invertors supplying electrical power to the gun-firing circuit when the undercarriage was down, to prevent the guns being fired while the aircraft was on the ground.

Swift FR.5 of 79 Squadron.

A Canberra PR.7 of 31 Squadron.

Lightning F.2A of 19 Squadron.

Wessex of 18 Squadron.

Whirlwind on a jungle pad in Borneo.

Air drop from a 48 Squadron Hastings.

Whirlwind heading for Brunei.

Phantom of 17 Squadron taxies at RAF Brüggen.

Vulcan being serviced at Gan.

Preparing for a 35 Squadron photograph.

Above left: A Beverley displays its huge rear loading doors.

Above right: Harriers en route to Norway take on fuel from a VC-10.

Below: Hunters of 8 Squadron at Khormaksar joined by two USAF B-57s.

Chinook of 18 Squadron.

An 8 Squadron Hunter FGA.9.

Above: Harriers of 4 Squadron over the Möhne Dam.

Left: 31 Squadron Tornado GR4.

Above: Chinook of 18 Squadron on a field exercise. (AHB)

Middle: Tornado GR4 over the Lake District, February 1990.

Below: The mighty Chinook.

As an additional precaution, a modification was introduced whereby a plug was inserted into the gun-firing circuit; known as the master armament safety break (MASB). Access to the plug was via a small panel under the port wing directly behind the engine air intake. It was the responsibility of an armourer to stand by the peri-track and disconnect the MASBs on each aircraft as they taxied in and connect them when taxiing out. An instruction was also issued that an armed Hunter must be identified by the placing of a red-painted ammunition box in front of its nose, attached via a short bungy cord to the top of the nosewheel door. Personnel were also forbidden to walk in front of an aircraft identified as being armed.

TIME TO RELAX

After an afternoon nap and with the heat of the day relenting, it was off to Steamer Point to haggle with shopkeepers in the Crescent, take a dip in the pool or make for one of the shark-protected beaches, a game of tennis, football or rugby for the energetically inclined, before winding down with a can or more of 'Slops' (Allsops lager) in the Camel Club.

... NOT FOR LONG!

That easy-going routine permanently changed in September 1962 when, following repeated strafing attacks by Yemeni air force MiG-15s on villages close to the Yemen border, the

The Khormaksar football team with Ray Deacon front row second from right.

Hunter squadrons were tasked to fly dawn to dusk patrols along the 300-mile-long border to deter further incursions. As one of the two squadrons was on the rotating detachment in Bahrain, it was left to the squadron at Khormaksar to undertake the border vigil. This involved a pair of Hunters taking off at dawn and flying north-east along the coast before turning west to patrol the mountainous, indistinct border before returning to Khormaksar from the north-west. Pilots and ground crews were assigned to one of two shifts, the first starting at 0600 and finishing at 1300 with the second overlapping from 1200 to 1900.

Further pairs took off every sixty minutes to ensure that a pair was always in the vicinity of the border. Their gun packs armed and equipped with long-range, 230-gallon underwing tanks, a round trip took some one-and-a-half hours. Additional Hunters were maintained on armed standby at Khormaksar for upcountry army support.

Pressure on both air and ground crews during this period was intense, the high workload taking its toll on both man and machine. Aircraft unserviceability levels rose to critical levels due to fifty per cent of airmen being off-shift and the extreme climatic conditions. Welcome relief came in March 1963 when 43 Squadron was transferred from Cyprus to Khormaksar. The Tac Wing Hunter allocation now numbered forty-six aircraft, a formidable force with which to defend and police this vast area. Khormaksar had become the largest and busiest airfield in the RAF, hosting no fewer than nine squadrons and three flights. With three Hunter squadrons now on MEC strength, Operation Longstop could be restructured; thereafter, each squadron undertook the two-month detachment at four-monthly intervals.

A DREAM COME TRUE!

Since my arrival, not a single airman had enjoyed a trip in a T.7. To redress this anomaly, fellow airman 'Taff' John, and I took advantage of an alcoholic cloud hanging over a squadron bash in Bahrain, to badger OC 8 Squadron, Squadron Leader Tam Syme, into giving airmen the experience flying in a T.7. He said he would bear it in mind, and he did not let us down. Shortly after returning to Khormaksar, he introduced a policy of offering airmen flights in a trainer within a couple of months of their becoming tour-ex.

In January 1964, 'Taff' became the first 8 Squadron airman to savour the delights of a trip in a Hunter. Then it was my turn, and I finally realised my dream of flying in a T.7. With Sid Bottom at the controls, XL613-Z took to the clear blue skies in the late afternoon of 14 February. Keeping low, we flew north-east along the sandy coastline, passing a customs post and isolated village, before climbing to 10,000 ft for some aerobatics. Sid showed me how to loop and barrel roll before letting me loose. I had been taught how to do both on the Vampire T.11 by Pete Broughton, a QFI at CFS, so I had some idea of what was required. My efforts were not so successful as with the T.11, but it was fun and I enjoyed having a go.

Sid then suggested we did some 'Wadi-bashing', which sounded exciting. So, we headed north into the mountains where peaks reach 7,000 ft or more. My 'driver' knew the rocky escarpment like the back of his hand, and we hugged the sandy valley floors, flying at around 500 knots, the steeply sloping cliffs rising high on either side. I suspect the roar of the Avon would have echoed long after we had gone. I had been quietly hoping for a sedate ride to take photographs and on enquiring as to why we were flying so fast, exciting as it was, Sid calmly replied, "to avoid being shot at". An illuminating answer which had me looking for tribesmen perched on the rocks as they flashed by above us. The turbulence prevented the taking of photographs, so I settled back and watched in awe as we weaved around the wadi floors. One memorable image that appeared through the windscreen comprised a narrow, u-shaped rock formation jutting out from the Wadi sides like a natural wall, the narrow gap at the base allowing sufficient space for the Hunter to pass. The valley floor on the other side dropped away a thousand feet or more, so down we went to continue this thrilling, high-speed ride. In a blink of an eye, we were out into the open, flying low and fast over a flat landscape on a southerly heading towards Khormaksar. I asked Sid if we could climb and fly round Shamsan so that I could take some photographs of Aden and he kindly obliged. To conclude one of the most exhilarating experiences of my life, we touched down and my head flopped forward as the brake-chute deployed and the brakes applied.

THE DESTRUCTION OF FORT HARIB

Following a rocket and bomb attack by Egyptian MiGs and helicopters on a guard post and village near Beihan, immediate retaliation was ordered by the British government on the Yemeni Fort at Harib, some three miles north of the border. At 0330 on 28 March 1964, 8 and 43 Squadrons ground crews were aroused by the bellowing voice of the duty sergeant, instructing us to report to our respective squadrons: no shaving or washing, just get dressed and board the transport waiting outside. Fortunately, the rare sobriety of the previous evening meant that hangovers were few as we were driven away. The realisation dawned that, having never been called out in the middle of the night before, something exceptional had happened.

On arriving at the squadron, our pilots could be seen excitedly darting between offices. No reason was given for disturbing our sleep as 'Chiefy' issued instructions to arm seven 8 Squadron FGA.9s (two in reserve) and two 1417 Flight FR.10s (one as reserve), as 43 Squadron personnel prepared five FGA.9s (one in reserve) for a joint Tac Wing operation. As soon as our pre-flight checks had been completed, we joined the riggers and armourers

The Khormaksar Wing's Hunter T.7.

in replacing the outer drop-tank pylons with rocket rails, a lengthy and fiddly undertaking on the poorly lit pan. Rocket trolleys were then towed behind the aircraft and, while some helped to mount and connect these highly explosive projectiles, others stuffed warning leaflets in the wing flaps of the two FR.10s and two FGA.9s.

As the sun came up, Peter Lewis (FR.10) and Sid Bottom (FGA.9) walked out to their respective aircraft, each equipped with HE gun packs and leaflets in the wing flaps. As they started up and taxied out, I was assigned to one of the eight FGA.9s tasked to carry out the strike. My pilot, Martin Johnson, was very excited and he eagerly completed his pre-flight checks, before running up the ladder and jumping into the cockpit. I handed him his parachute and seat straps, waited until he was safely belted in, removed the ejection seat safety pin, and stowed it away before removing the ladder. At a given signal from OC Tac Wing, Wing Commander John Jennings, in the lead aircraft, eight pilots pressed their starter buttons and seven Avpin starters burst into life, their Avon engines lighting up within a few seconds. Why seven? Sod's law had kicked in; the starter on Martin's aircraft failed. Having given several hefty thumps on the starter relay boxes with the handle of my screwdriver, I gave Martin the signal to try again. Silence!

In an instant, the straps were flying over Martin's shoulders, and he was standing on the seat, helmet still in place, ready for the ladder to be attached. Having clambered down, he sped to a spare aircraft with me in hot pursuit. As the seven Hunters began to taxi, he clambered in and between us, he was belted up in record time. Safety pin out, ladder away, a twirl of the forefinger and a welcoming blast from the Avpin starter. The engine could hardly have reached idling speed when he throttled up and raced out onto the taxiway in a desperate attempt to catch the others as they accelerated down the runway. He must have reached a similar speed as he made his way towards the threshold, round the corner and on to the runway with Avon seemingly at full power throughout. Off he went in hot pursuit of the seven as they disappeared into a distant haze.

Meanwhile, Sid Bottom in the 'solo' FGA.9 flew fast and low over the fort, lowering the flaps to release the warning leaflets as Peter Lewis in his FR.10 took photographs as evidence that a warning had been given. The Yemeni troops now had fifteen minutes to vacate the site or face the consequences. On arrival the main force went straight into the attack using both 30-mm HE cannons and rocket fire, leaving the two leaflet-carrying aircraft to maintain top cover. Once the smoke had cleared, Peter Lewis made a final sweep to photograph the damage inflicted on the target. These revealed that the stone-built fort was virtually destroyed, along with several anti-aircraft guns and vehicles. The attack had the desired effect, incursions by Yemeni aircraft ceased; at least for the time being.

TOUREX

All-in-all, with ten months in Bahrain, three weeks in Kenya to cover a rebellion by the country's army and a month on home leave, my two-year tour flew by. The camaraderie and family spirit on 8 Squadron made time in Aden a more bearable experience and, having survived the sun, heat, sand, and flies, I looked forward to the sanctity of life in the UK. Or not! After two years in a faraway land where the locals spoke foreign, the beer was just about tolerable and the climate unbearable, my posting came through – RAF Valley on Anglesey – where the locals spoke a foreign language, no beer on Sundays, and an iffy climate. What had I done wrong?

CHAPTER FIFTEEN

BEVERLEY NAVIGATOR IN ADEN

RAY HARPER

My tour of duty in the Middle East was in 1966 and 1967 up to the withdrawal of British Rule in Aden as British forces began withdrawing from East of Suez. A British United Airways VC10 took me from London to Aden where I joined 84 Squadron operating the huge Blackburn Beverley. Passengers new to the Middle East faced an outside temperature of 110°F when the cabin door was opened. One wag likened it to "sticking your head in an oven on Gas Mark 9". The VC10 was locally known as the 'Moon Rocket'; this because new arrivals from the UK had the pale skin of 'Moonies'. Acclimatised personnel were deeply suntanned, the contrast was immediately obvious.

We were briskly ushered to the arrival's hall into air-conditioned comfort where we were processed for our entry into Middle East Command. The bus that took us to Medium Range Transport (MRT) Wing HQ was air conditioned; the air was cool but the bus not capable of keeping fine dust and sand out of the cabin – a condition applied to several workplaces. From the bus we viewed a bland brown, grey, and concrete airfield. The landscape to the north displayed rocky desert and mountains known as the Jebels. RAF Khormaksar had the appearance of airfields I had seen portrayed in old war films that invariably had John Mills in the cast.

Ray Harper poses with his 84 Squadron Beverley.

Induction into Aden and my unit began after the initial perfunctory welcome. The personnel tasked to meet us had been called in during the heat of the afternoon, outside their normal working hours. Although the base was operational twenty-four hours a day, most work began before sunrise and ended in the early afternoon. The high afternoon temperature was cruel to man and beast.

My leader met me, showed me key places in the wing, then dropped me off at the officers' mess to check into my accommodation. He was clearly keen to leave and return home to his family; they were at the swimming pool. A bored local showed me to my large room that had a great view to the south from a first-floor terrace towards Shamsan, the mountain known as Crater. In the heat I thought I saw shimmering pink salt flats, an illusion caused by the heat and thousands of pink flamingos that migrated to the flats from Africa.

I met my 'batman', a tall Somali; he was aged about thirty. He stood and bowed. My Somali was minimal, his English poor but we learnt to communicate successfully. He insisted on unpacking my baggage and ushered me out of my room, I guess to let him work in peace. In the cool of the mess lounge, I spent a couple of hours reading my briefing notes. After freshening up I stood on my terrace to watch the sunset, a rapid transition from day to night, typical in these latitudes. Sounds of laughter and conversation from the outside bar – known as the Jungle Bar – no jungle was anywhere in sight – drew me to join the large gathering of officers and ladies who were enjoying their first drinks of the evening. Introductions were made. As a Moonie there was no doubt that I was a new boy. I met a few guys who were on 84 Squadron and was invited to join them at dinner.

The dining room was massive with no air conditioning but served by several ceiling and wall fans. Coolish evening air flowed through several open doors. A doctor, dining at my table, explained the need for the regular intake of fluids to avoid dehydration and other related issues. He pointed to large jugs of orange and lemon juice – known as 'Jungle Juice'. It was laced with salt, an essential ingredient in hot climates. He told me that once I started to taste the salt then I had consumed enough.

During my first two weeks on 84 Squadron, I met the boss and the section leaders, flight commanders, and the engineers, and admin staff. I visited the other squadrons on MRT Wing; 21 and 105 Squadrons. Visits to the other units – fighter, coastal, helicopter, and recce squadrons – made me realise that this really was an air force in its own right. The missing element was a bomber squadron, but the region was reinforced with Canberras from RAF Akrotiri in Cyprus or from the Far East. Desert and sea-survival training was stunningly realistic with the emphasise on combat survival training. The RAF Regiment provided weapon training and live shooting practice, and it turned out that I was a good shot.

The squadron training staffs, a navigator and pilot, provided a detailed brief on upcountry airfields and strips that were used by the Beverley. They highlighted regions of high ground that posed significant problems, particularly in bad weather, sandstorms etc. Specific trouble spots underlined where enemy activity could be expected. The squadron route structure was outlined, which encompassed the Persian Gulf, in particular Kuwait, Bahrain, Sharjah, and destinations within the Trucial States. There was also Muscat and Oman, with a main base on Masirah Island, and staging airfields at Salalah, and Riyan. Heading south from Aden, routes took us to Nairobi and Mombasa in Kenya, and South Africa with a few sorties extending to Rhodesia. Occasionally, the squadron headed for India, and as far as Hong Kong. Command staff briefed us on the political and security situation. At the time of my arrival, Aden was a dangerous place. The National Liberation Front (NLF) wanted the British out as did the Front for the Liberation of Occupied South Yemen (FLOSY). Both parties' military units were willing to cause mayhem, death, and destruction to achieve their aim.

My first flight was a local familiarisation trip with a screen navigator outlining points of interest and danger areas along with locations where RAF aircraft had been fired upon. We dropped supplies by parachute onto a drop zone (DZ). My 'screen' confirmed my tactical airdrop qualifications enabling me to start operational flying, and I expected to start upcountry operations immediately. However, my first full month's flying included a

An 84 Squadron Beverley lands at the forward airstrip at Thumier, Aden.

recce task to Socotra Island (Horn of Africa), three tasks to Nairobi in Kenya (turned back on the second trip because of an engine failure), a flight to Salalah and Bahrain, and a flight to Salalah and Sharjah (turned back to Sharjah with a technical problem). In less than three weeks my logbook listed over eighty hours day and night flying. Squadron training and examining staff had assessed my skills and authorised me to fly on operations unsupervised.

The bulk of Beverley flying supported British and local forces in the Aden Protectorate, and in Muscat and Oman. Two upcountry airfields had paved runways, Beihan and Mukeras. Beihan was in a wadi orientated north–south with mountains either side of the narrow valley. Mukeras sat at 6,000 feet above sea level on a jebel. All other Beverley destinations in the region were sand and rock strips, some very rough. A few notorious strips lay close to the Yemen border when active Yemeni units often fired on the strips from mountain ridges. During my tour several aircraft returned to Khormaksar holed by heavy-calibre machine-gun and rifle fire. The Radfan was a particularly unfriendly, and sometimes deadly, region. A notable 84 Squadron casualty was Beverley XM106 (see photo overleaf). During a dawn assault, it ran over a landmine placed overnight on the 'secure defended strip'. The aircraft was blown up by army engineers after the few valuable items were removed for recovery to Aden. During the last few years of British occupation of the Protectorate, a few other aircraft of various types were lost to enemy action or as the result of bad weather around difficult mountainous terrain and landing strips.

Several operations involved combined resources drawn from the other theatre squadrons, sometimes augmented by Cyprus-based Canberras on detachment. On several occasions I witnessed Hunters in the ground-attack and close air-support role. I watched a Shackleton open fire on a supply caravan consisting of camels and vehicles. A line of camels laden with supplies exploded in a spectacular fashion, synchronous blasts running along the supply line, the blasts removing several of the vehicles.

Social life on the squadron, the station, and at the beach club provided essential relaxation and the opportunity to meet and enjoy the company of service personnel, their families, and civilians, most of whom were keen to party. Squadron and wing events were frequent, drawing together a wide range of people from most units on the huge base. These gave us the chance to talk directly to crews of the Argosy, Twin Pioneer, and Dakota squadrons that made up MRT Wing, together with the Strike Wing Hunter pilots of 8 and 43 Squadrons, and the 37 Squadron Shackleton crews. The squadron commander ensured that our 'beer calls' included army, Royal Marines, and civilians who we worked alongside. Families were entitled to take leave in Kenya at a variety of locations, game parks, and Nyali beach near Mombasa. Many took advantage of the popular list of leave options. On a couple of occasions, I enjoyed being a part of a crew that flew families to Kenya for their holidays.

The sabotaged Beverley at Habilam Strip.

As time passed, the security situation in Aden deteriorated. During a visit to a squadron family, who lived downtown in an apartment on Maalla Strait, they took me under their care. While John prepared the barbecue, and his wife and daughters prepared the meal, they kept me fed with beers when we heard a loud commotion outside in the street. A rocket-propelled grenade came through the wall of the lounge, passed across the room, and smashed its way back outside, without exploding. After the initial shock, we sheltered in the central stairwell and waited until the security forces removed the threat. Over the next few weeks similar incidents occurred.

In 1967 the security risk to families was assessed as high, leading to them being returned to the UK or posted to other locations. In a short period of time, the manner and style of life in Aden changed radically. Most military units withdrew to defended bases, such as RAF Khormaksar and HQ Steamer Point. The days of colonial living in peace and comfort were over. Married men who had waved farewell to their wives and children moved into base accommodation. Men shared rooms in a variety of accommodations, messes, barrack blocks, apartments. The pace of squadron life increased with much more flying upcountry, often in difficult circumstances. Each crew member was 'armed to the teeth'. Additional shared work included guard and patrol duties. Transport aircrew and support staff had the benefit of getting away from Aden to less stressful destinations. Flights around the Command included aero-medical, evacuation, resupply and supporting rotating squadrons around the Middle East and around the Near East.

Beverley and Argosy aircraft flew Combined Services Entertainment (CSE) performers around Aden and the Middle East. A most popular and welcome distraction for us who were in the danger zone. We applauded the entertainers for their courage and commitment to our troops. Familiar names such as Harry Secombe, Bob Monkhouse, Mike and Bernie Winters and others. However, the lads were more interested in the many attractive dancers in the CSE line-up. I got to know a few of the entertainers better while I was part of a Beverley crew that flew them to Riyan, Salalah, Masirah, Sharjah and Bahrain. However, we felt sympathy for many of the artists who succumbed to 'Montezuma's Revenge' because they had not acclimatised to the food and were struck down by local illnesses.

MRT Wing was tasked to rotate army regiments in the Wadi Hadhramaut that lay about 400 miles north-east of Aden. After landing on Ghuraf, a sand strip, the Irish Guards de-planed and dispersed. The sheik of a nearby town met the British commanders and drove them to the city of Shibam to enjoy his hospitality. Luck was with me because I was included amongst the people loaded into four Land Rovers. Shibam was a city built against the side of the Wadi. It boasted early skyscrapers that dated back over 2,000 years. The richness of the soil, watered by irrigation ditches, was verdant and colourful with fruits and crops – an unexpected sight in the middle of the Arabian Desert. During our flight back to Khormaksar, we watched four Hunters strafe and rocket caves that held insurgents who attacked our supply lines.

Beverley, Argosy, and Twin Pioneer upcountry tasks involved landing to offload and upload troops and supplies. Air drops were rare and involved lightweight packs to resupply deployed troops. Yemeni forces made several attempts to disrupt our supply line and attacked when they saw the opportunity. We gained support from the two Hunter squadrons. They flew at high altitude overhead the airfields and strips ready to descend to provide firepower. UHF and VHF frequencies were set and defined as 'Black spot and Blue spot'. Our signallers called for help when ordered by the captain. A Hunter would overfly us and the insurgents at very low level and at high speed. Exciting, startling and, for us, most welcome. When the enemy were unimpressed by a low-flying demonstration, the Hunters occasionally used their 30-mm Aden cannon to reinforce their determination, also most welcome from our point of view.

A Beverley task to Beihan was to extract an Arab clan that was under threat of death from local tribesmen. While boarding the passengers, an ugly group of armed men approached us obviously keen to do harm. Blue spot was used to call in air support. To our surprise, rather than a Hunter appearing, we watched a Shackleton approach from the north at very low level. As it passed abeam us, the gunner opened fire from the forward turret, letting loose a fusillade from his guns that had an instant impact on the insurgents; they scattered and ran off into the desert.

On 7 March 1967, 84 and 30 Squadrons were on operations onto Socotra Island to dislodge communist-backed guerrilla forces intent on removing Omani authority. Three

Beverleys were employed to fly the Hadhrami Bedouin Legion under the command of an Omani-employed British colonel. I was on the first Beverley that landed at a disused wartime airstrip on Socotra at dawn. As the second and third aircraft approached Socotra, the battle at Hadibo Fort was won by the Legion, and the two airborne Beverleys ordered to return to Riyan. My aircraft landed at Riyan after an eleven-hour mission.

In July 1967 an industrial strike at Aden's harbour substantially reduced supplies with a reduction in aviation fuel a significant problem. Our Beverleys, with outer wing fuel tanks blanked off, were flown to Assab in Ethiopia to pick up AVJET fuel in these outboard tanks to provide fuel for the Hunters back in Aden. During these tanking tasks, the opportunity was used to airlift stocks of goods to Aden, which included hundreds of thousands of cigarettes, beer, and spirits. Ours was a most popular contribution to the garrison and its military and civilian personnel. We, the unglamorous truckies, took pleasure in providing fuel for the Hunters, as well as the keenly demanded provisions for the troops, sailors, and airmen.

On 1 August 1967, I flew back to the UK to get married; hoping that I would not need to return to Aden. However, I received a telegram ordering me to return by 18 August. When I got off the BOAC VC10 at Khormaksar, I was met by the squadron hierarchy. It was clear to me that something was amiss; my first thought was about my new wife. They informed me that a friend and air quartermaster on 84 Squadron had been assassinated downtown on Maalla Strait while refuelling his car. There was much anger in the air.

A Beverley creates its own sandstorm at an upcountry strip.

Ray Harper and his crew prepare to leave Khormaksar for the final time.

In September, the flying task took me to Assab, Jeddah, Sharjah and Salalah as well as to several upcountry strips. Contraction of British forces in Aden was now accelerating. Three of the last six Beverleys returned to the UK routing through Bahrain, overflight of Iran to Diyarbakir in Turkey, then to Cyprus, Malta with Abingdon in Oxfordshire as the destination. October saw the heavy burden of the final withdrawal from upcountry impact on 84 Squadron. Riyan Mukalla airport was evacuated by British Forces on 16 October. My Beverley departed Mukalla with a heavy payload. The task was extended for us to provide a radio relay for the Hunters as they provided air defence for the withdrawal from Mukalla. On 27 October 1967, I believe I was on the last upcountry task from Beihan to Khormaksar with passengers who included the local diplomatic officer, rear-guard troops, and valuable airfield equipment. It was a full aircraft and a heavy load.

On 29 October, a fly-past over Khormaksar, including RAF and Fleet Air Arm aircraft, saluted the AOC-in-C AFME. Wessex helicopters, the 21 Squadron Dakota and 84 Squadron Beverleys, followed by Hunters, Sea Vixens and Buccaneers flew over in a timed order with the aircraft heights staggered with the helicopters at a low height with the Fleet Air Arm contingent bringing up the rear at a higher level. I was in the lead Beverley so could see the entire formation from my astrodome. By this stage, all other squadrons and flights had finally departed Aden for other airfields in the Persian Gulf. The main base at RAF Khormaksar looked empty.

On 6 November 1967, the last three Beverleys departed RAF Khormaksar in Aden to return to the UK. The return flight is a story in itself. A tale of very bad weather, diversions, engine failures, and an attack by Egyptian fighters on one of the Beverleys. Royal Navy aircraft carriers returning to the UK from the Far East carried personnel and equipment out of Aden, ending the years of British colonial rule in the Aden Protectorate.

CHAPTER SIXTEEN

WITH 208 SQUADRON IN BAHRAIN

CHRIS GRANVILLE-WHITE

After three years at the RAF College Cranwell, followed by advanced flying training on the Vampire T.11 at Valley (as the promised Gnats had not arrived on time) there was the Hunter OCU course at Chivenor to relish. Our course comprised John 'CJ' Thomson (later Air Chief Marshal), Martin Herring, Tony Wilcocks, and me.

From Chivenor, I joined 208 Squadron at RAF Khormaksar (Aden) on 25 June 1963, a week before my 22nd birthday, for a two-year posting. Aden was a barren place and not very pleasant, although as it turned out much more interesting than Bahrain where we were destined to move a year later. When I told my father (a retired air vice-marshal and World War One pilot) that I was posted to Aden he advised me to see his solicitor to make my will.

The unofficial policy on 208 Squadron was to fly as low as we could always, except when the boss was in the formation. The boss made it clear that any pilot flying below 250 ft AGL would be sent home. This sometimes felt like an attractive option, but hardly a good

No. 208 Squadron at RAF Muharraq. Chris Granville-White is standing on the extreme left.

A 208 Squadron FGA.9 over Masirah.

career move. With no radio altimeter or low-height warner in those days, it was difficult to judge the height when flying over the rocky desert with few features to give any indication of how high you were.

After I had been on the squadron for two months, it was our turn to be detached to the Persian Gulf (since re-named the Arabian Gulf) for two months on Operation Longstop. Following the Iraq invasion of Kuwait in 1961 (to be repeated twenty-nine years later by Saddam Hussein), one of the three Khormaksar Hunter squadrons was always at RAF Muharraq in Bahrain. This was arranged on a two-monthly rotation so each squadron would in turn spend four months at Khormaksar and two months in Bahrain – including a two-week detachment to Sharjah for armament practice camp.

The role of the detached Hunter squadron in Bahrain was for the protection of Kuwait if Iraq might decide once again to claim ownership of this small oil-rich state, which was their rightful territory – as a cursory look at the map shows that it could be a logical conclusion. We never flew the Hunters north of Bahrain in the direction of Kuwait, presumably to avoid antagonizing the Iraqis.

In April 1964, nine months after joining 208 Squadron, I returned to the UK for a routine one-week simulator training course at Chivenor. While I was away, the squadron officers were dined out in the officers' mess at Khormaksar during which the boss in his speech mentioned that he had heard about the move to Bahrain from his wife who had heard the news at the beach. Apparently, not long after this comment was made at the dinner, one of the staff officers from the AFME HQ at Steamer Point slipped away from the dinner and an admin order for the move appeared at the squadron next morning.

Skip bombing over the weapons range.

TO BAHRAIN

The other squadrons based at RAF Bahrain were No. 30, which had recently moved from Nairobi with their Beverleys, and No. 152 equipped with a mix of twin-engine Pembroke light-transport aircraft and the slow-flying short-take-off-and-landing Twin Pioneer light-transport aircraft known as 'Twinpins'. The airfield was also the civil airport and located on the small island of Muharraq, connected to the main island by a causeway leading to the capital Manama. The following year RAF Bahrain was re-named as RAF Muharraq.

The officers' mess was a small, white-painted single-storey air-conditioned building by the sea. This may sound idyllic, but the reality was that as the tide went out a long way it revealed a vast expanse of stinking seaweed (and goodness knows what else) so that life in the mess was only made bearable by the air conditioning which kept this awful smell at bay. It certainly wasn't a beach for swimming.

The station commander who took over at about this time was Group Captain Brian Frow, a decorated wartime bomber pilot, who lived near the mess with his vivacious wife and four teenage daughters. They were a popular family who always welcomed us young pilots, even when we turned up uninvited. Indeed, one of our pilots was later to marry one of the daughters. Wing Commander Fitzgerald commanded flying wing and on a few occasions some of us joined his family for a weekend day out on a dhow for a picnic and swim, while the local crew cooked their own meal over a charcoal fire on the deck – a questionable practice on a small wooden vessel.

Although the move to Bahrain was popular with the married officers and ground crew, who no longer had to leave their wives alone in Aden during the two-month detachments to Bahrain, it was a major disappointment to me. The flying in Aden had been interesting and varied, but flying from Bahrain was much less interesting and we missed out on the continuing Radfan operations in Aden. Bahrain as a place to live was boring in the extreme with very little of interest on either the main island with its capital and the few villages, or on the tiny island of Muharraq. There was an excellent swimming pool at Muharraq, but no decent beach and few single girls. To fill in some of my free time during that long year I completed a market gardening correspondence course which provided a fascinating and pleasing change from the daily life in the colourless backdrop of Bahrain. With Bahrain's location further north of the equator than Aden it was a few degrees hotter in summer and cooler in winter.

BAHRAIN FLYING

Having moved to Bahrain, we spent about two-thirds of the following year at Muharraq and one-third on detachments to Sharjah. When at Muharraq the main flying activities were air-combat training over Qatar; four-ship hi-lo-hi strike or recce sorties to the Trucial coast; and naval cooperation in the Gulf with the Tribal class frigates HMS *Gurkha, Mohawk,* and *Zulu.*

The hi-lo-hi strike or recce sorties from Muharraq were mostly to the coastal and inland areas south of Sharjah, Tarif and Jebel Dhana. These sorties were often bounced by a 'menace' Hunter flying from Sharjah, and I did my share of such flying as a singleton. The naval cooperation involved four-aircraft strike sorties when we tried to outwit the ship's radar to make a simulated attack while they fired break-up shot from their huge guns. This shot only travels about 200 ft so we could still get reasonably close before breaking off the attack. We also used our 30-mm cannons to strafe a splash target towed behind the ships. Each of these frigates visited Bahrain from time to time and we came to know some of the crews.

Once a month, to remain 'current', we flew night sorties from Bahrain to Sharjah and back. As our only night role was to be capable of deploying to another RAF airfield, night flying involved aircraft setting off at ten-minute intervals to fly high level to Sharjah (264 nm each way) then let down and make a low overshoot of the runway before climbing back to high level for the return to Bahrain. Some of us might do a few barrel rolls during the transit across the Gulf but I gave this up after becoming somewhat disorientated one night while trying to work out which way was up in relation to the only visible landmark of the flare stack burning at the oil refinery on Das Island – a single bright light off to one side does not provide a good visual cue for orientation. On one occasion it was decided to have a competition in which we would each record after landing our best guess at the title of the film being shown in the open-air cinema at Sharjah. We would only be able to achieve a very brief glance at the screen as we flew over and it must have been very annoying for those watching the film when every ten minutes a Hunter would fly near the cinema at slow speed with high power – with hindsight perhaps not a very wise procedure.

MID-AIR COLLISION

Our permanent time in Bahrain began badly. Tony Wilcocks had been back in the UK to collect his wife Jill and they had flown out to Bahrain to move into a flat to resume married life after a year apart. Sadly, this period was to last only a few days because at the end of their first week together Tony was killed in a mid-air collision with Mike Gibson who ejected safely. Tony was leading his first tactical four-ship ground-attack sortie against a simulated target near Yas Island just off the Trucial coast near Jebel Dhana. I was in the crewroom when the message came through that there had been a mid-air collision and that two pilots were missing. The formation led by Tony comprised Sandy Aitken as No. 2, Mike Gibson as No. 3 and Frank Mitchell as No. 4. Apparently, Tony had picked up the target late, so that instead of the four aircraft attacking in the correct sequence, with the planned time interval between them, he turned in late which put him on a collision course with No. 3 and they collided over the target. Mike climbed up to about 6,000 ft in his damaged aircraft and ejected. Frank Mitchell took gun-camera cine of the area of sea where they collided in the hope of spotting Tony, and after circling for a time Frank and Sandy returned to Muharraq and informed air traffic control about the accident. The boss went to tell the two wives what had happened, and they stayed together awaiting news. Later in the day a message came through that a pilot had been picked up by a passing dhow and dropped off near Jebel Dhana. It must have been hard for Gordon Lewis to have to re-visit the two wives to tell them that Mike was OK and Tony had not survived. A Pembroke of 152 Squadron was soon on its way to collect Mike. When it landed back at Muharraq later that afternoon a few of us went to see a very sunburned Mike as he was lifted out on a stretcher because of the usual fractured vertebrae from the ejection.

Tony's funeral took place a few days later. Anthony Mumford nominated Tim Webb, Keith Blatchford and me (the tallest pilots) to be the pall-bearers with him and after he had drilled us in the procedures, we went off in an RAF ambulance to collect the coffin from the morgue, which comprised a surface with a few Arab bodies lying there and Tony's coffin marked with a label, which we carried out to the ambulance and continued to the Roman Catholic church. The service was conducted in Latin, so that the only words we understood were Tony's name after which we lowered him into the shallow grave outside. This was a dreadful tragedy, but sadly not unknown in RAF service.

DANGEROUS DINGHY DRILL

One morning in August 1964, after a very late party the previous evening, the boss announced that none of us was fit to fly so we would be going out on the RAF air-sea rescue launch instead. We boarded the launch together with a good supply of beer and headed off at speed into the Gulf. Once we were well out of sight of land the launch slowed and the boss said that the three youngest pilots would be doing dinghy drills. Keith Blatchford, Jock Watson and I duly put on a parachute harness each with the dinghy pack and jumped off the moving launch which then sped away with the other pilots waving their beer cans at us. Dinghy drills were a familiar procedure from the Chivenor course when going to sea in the Appledore lifeboat, jumping overboard and later being winched up by helicopter. Once the three of us were in our dinghies that day in the Gulf we tied the dinghies together and switched on the one 'SARAH' (SARBE predecessor) rescue beacon we had been given which would show our location to the rescue launch. By now the launch was out of sight and as time went on it became extremely unpleasant sitting in our dinghies under the August sun. Jock had only joined the squadron on the previous day and after a very late-night party in London before catching his flight to Bahrain, followed by the late party soon after his arrival, he was being sick and becoming very dehydrated as we had no fresh water. We waited and waited – this was no longer funny. Eventually the rescue launch reappeared with some anxious faces on board. It turned out that the SARAH had been used many times for training and not long after we had been dropped off the battery was exhausted. As the flight safety posters remind us, 'Don't assume, check!' Had the rescue launch not found us when it did, the result for all of us (Jock Watson in particular) could have been very serious.

MASIRAH FLYING ACCIDENT

During October 1964, a recently arrived first-tour pilot at Khormaksar, Flying Officer Ian Stephens, was killed shortly after taking off from Masirah. This was at the end of a short detachment with a combined team from 8 and 43 Squadrons. Shortly after take-off the six Hunters were joining up for the return to Aden when the accident occurred. Ian's flight commander told me later that he was about to warn Ian that he was getting too low when the aircraft went into the sea, emerged briefly, and broke up.

ANTI-BRITISH DEMONSTRATIONS IN BAHRAIN

During March 1965 there were some anti-British demonstrations in Bahrain. I was the station duty officer at Muharraq when the first demonstration began one evening. I was contacted by the duty sergeant in the guardroom and went on my Honda to see what was going on. Outside the main gate a group of protesters were shouting and throwing stones and the small RAF duty guard were standing by with dustbin lids as protective shields. The situation was not too serious and after a while the crowd dispersed. But a few days later, school children and their teachers assembled on the causeway and threw stones at passing British vehicles driven by servicemen and civilians on their way to work. This was a frightening and potentially dangerous situation with some windscreens broken. RAF buses and Land Rovers were soon fitted with metal mesh protection for windscreens, for transporting personnel and families from their flats on Bahrain Island and across the causeway to Muharraq. This worked well for the first day or two, but the protesters then responded by throwing paint at the passing vehicles so that the drivers could not see where they were going, and attacks could then be made against the side windows. Another of their tactics was to nail bottle tops onto the road surface with a second nail pointing upwards to puncture tyres. Although these attacks only lasted through March it was extremely unpleasant and frightening. For those of us living in RAF accommodation at the airfield it had little effect on our daily lives however, and I had a busy month of flying.

SHARJAH FLYING AND TRAVEL

During the detachments to Sharjah the main activities were sorties to Rashid Range firing rockets and guns, and FAC training and courses with the army at inland locations. The periodic two-week detachments were more interesting than being in Bahrain, with occasional opportunities to travel inland. One weekend, some of our ground crew had arranged to go camping and booked a RAF three-ton truck and Land Rover plus an interpreter. They were told that they could only go if there was an officer with them and the ground crew asked me if I would join them.

We had a fascinating drive to the north-east up the coast to Ras al-Khaimah where an impromptu football match was soon arranged with a local team. We played wearing the ubiquitous 'noddy boots' and they played bare foot – and beat us easily. It was a hard game but a great way of making local contact and, after the match, those of us who had played were led to a nearby communal open-air stone bath to soak under a roof structure for sun protection. We camped on the beach for the night and kept a lookout during the hours of darkness because at the foot of the nearby mountains, which rose steeply a few hundred yards away, were caves occupied by the local Shih Hua tribe which had a fearsome reputation. Fortunately, they left us alone, although when we drove away next morning they came down to the beach where we had camped and dug up the empty food tins which had been buried.

On a few occasions at Sharjah, I was able to go away for a weekend with British soldiers of the Trucial Oman Scouts (led by a corporal) in a pair of Ferret Scout cars. During these trips we once went through the picturesque, rugged mountains towards Khor Fakkan. I was also able to arrange the use of two RAF Land Rovers and a local guide for a few of us from the squadron to drive to the Buraimi oasis on the border between the Trucial States (now UAE) and Muscat and Oman (now Oman). It was a long drive across loose sand dunes, with a twisting pass over an area of rising ground known by its curves as the 'Pass of the Neck of a Camel'. This journey took several hours bouncing along in the hot and dusty vehicle, but today it would be a short drive on a dual carriageway in an air-conditioned car. During the latter part of this journey, I became sick and was left at the British Army fort at Buraimi to recover. An army medic corporal at the fort gave me a jab in the backside, and after a couple of days in bed a Twin-Pioneer of 152 Squadron flew out to pick me up from their desert strip to return to Sharjah.

FLYING FROM MASIRAH

During September 1964, we spent a few days at Masirah Island flying low-level recce sorties around Oman with its majestic Jebel Akhdar Mountain which peaks at 9,780 ft. By now I had moved up the squadron pecking order sufficiently, to have my own aircraft (J for Juliet) with my name and flying officer rank painted on the fuselage by the cockpit – we flew together for much of this detachment and whenever possible thereafter.

Chris Granville-White at the spartan RAF station at Sharjah.

TRI-SERVICE EXERCISE ON YAS ISLAND

One week during May 1965, I was sent on a tri-service exercise to Yas Island near the coast at Jebel Dhana. I embarked on a Royal Navy landing ship at Jufair naval jetty in Bahrain and, after two interesting and enjoyable days at sea, we were landed at a beach on Yas Island. Yas was then a small, deserted island a few miles wide. Once we had come ashore with the army units, they moved across the island to set up their training exercise and my four airmen and I set up an air-to-ground range on the north-east of the island. Once the exercise began next day, the Hunters flew in at high level from Muharraq, before letting down to low level to attack the range targets, with me acting as the forward air controller, before climbing back up to high level to return to Bahrain. The first Hunter each morning would drop a mail bag for my small range team. The bag was jammed into the closed air brake with the risk that if the airbrake was used by mistake during the descent from high level our mail would be lost in the Gulf. Fortunately, this never happened, and each morning we stood in a row to spot the drop which was carried out at a low speed of about 200 kts to retain the mail in the bag.

UK FERRY FLIGHT HOME

At the end of my tour on 208 Squadron, I was tasked to lead three Hunters back to the maintenance unit at RAF St Athan for major service (two from 208 Squadron and

Hunter FGA.9 makes a low pass.

one from 43 Squadron). As a pre-ferry fuel-flow air test with the additional 100-gallon fuel tanks outboard of the 230-gallon tanks, I flew on a hi-lo-hi flight from Muharraq to Sharjah and Tarif and back. On 26 June, my formation of three took off as a vic formation from Muharrraq for the leg to Tehran (one hour twenty minutes). It was fascinating, at last, to fly to the north end of the Gulf, passing to the east of Kuwait, and onwards to the NNE across the mountainous area towards Tehran. In particular, I enjoyed the descent to Tehran with a sequence of beacons to pass over at reducing heights for a visual approach to Tehran Mehrabad airport.

For the onward journey to RAF Akrotiri that afternoon, the take-off run seemed to last forever. With the combination of the weight of the full 230-gallon and 100-gallon drop tanks, and the 4,000 ft elevation of the airport reducing engine thrust, I began to wonder whether we would ever become airborne. But we weren't the first Hunters to pass this way, and we eventually staggered into the air towards the continuing mountainous area beyond.

This second leg took us from Tehran north-westwards towards the northern part of Iran, passing over the huge Lake Rizaiyeh and before long into Turkish airspace and the welcome landmark of Lake Van from where we headed west. Along this route towards Turkish airspace, the Soviet border was to the north of us with Mount Ararat visible in the distance. During this time of the Soviet era, some of the navigation beacons beyond the border operated on the same frequency (but at greater power) than the ones we were following, in a devious attempt to lure western aircraft across the border – termed meaconing. For this reason, we were required to map-read across this area and only to continue if there was less than half-cloud cover.

I called Tabriz Flight Follow in northern Turkey, but they did not provide any service. We continued westwards routing over the beacons at Elazig, Malatya, and Kayseri, before altering heading to the south-west to Adana and across the northern Mediterranean to Cyprus, where we landed at Akrotiri in the British Sovereign Base Area after a flight of two hours thirty minutes.

While at Akrotiri I overheard someone complaining about the hot temperature in Cyprus, which seemed laughable to us as the temperature was at least twenty degrees less than in the Gulf area. By now, the 43 Squadron aircraft had a technical unserviceability which couldn't be put right speedily, so it was left behind with the pilot for rectification. The next morning, my wingman and I continued our way, flying at high level westwards along the Mediterranean to RAF Luqa on Malta.

After a night stop at Luqa we set off for the long flight to the UK, heading north-west over the Mediterranean, passing Sicily and Sardinia, then northwards over France to the UK and the maintenance unit at RAF St Athan in South Wales. However, by the time we were over northern France we were too low on fuel to reach St Athan; so, when mid-Channel and leaving French airspace, I called UK Southern Radar for a diversion to Thorney Island on the south coast. By now my wingman's radio had failed so I waggled my wings to bring him into close formation before we descended through cloud. However, when I came out of cloud near Thorney for a visual join there was no sign of my wingman who had lost me in cloud. Being low on fuel there was nothing I could do about this, and I landed at Thorney after two hours forty-five minutes. Fortunately, my wingman soon found his bearings and landed a short time after me.

Thorney Island was then a Transport Command station operating Beverleys. The Wing Commander OC Eng Wing came out see our diverted aircraft. He was shocked by their sand-blasted appearance and after looking in the cockpit of my aircraft said that he would impound our aircraft as they were clearly unfit to fly. I explained that all the Hunters in the Middle East were in this condition, and we had just flown them all the way from Bahrain. Moreover, we only had a short distance to fly to St Athan and I asked him what he was going to do with the aircraft if we left them there. It was evident that he would be only too relieved to see the back of us and after the radio on the other aircraft was fixed, we flew on to the maintenance unit at St Athan in thirty minutes.

AFTERNOTE

I had very much enjoyed leading this fascinating ferry flight via the 'Northern Route' over the mountains of Iran and Turkey, which was a contrast to the 'Southern Route' from UK to Aden which I had flown as a wingman early in the tour. On that occasion, we had met up at Lynham, after collecting our respective aircraft from the maintenance units at St Athan and Kemble. From there we routed via RAF Luqa (Malta) then RAF El Adem (Libya) for a night stop; before continuing next day via a refuelling stop at Khartoum, to Khormaksar. The leg from El Adem to Khartoum had been challenging, with much time out of radio contact with ground stations despite the excellent range of the VHF at high level, no ground nav aids, and the need to alter heading before a lone landmark of an area of black rocks from an ancient volcano at the south-west corner of Egypt (known as Nasser's Corner, after the Egyptian president) for the route onwards to Khartoum.

CHAPTER SEVENTEEN

HELICOPTERS OVER THE JEBEL

TIM JENNER

ALTHOUGH IT WAS JANUARY, AND therefore the coolest time of year in Sharjah, descending the somewhat rickety steps from the VC10 into the humid evening air was a bit of a shock having left a snowbound Brize Norton earlier that day. However, not much can alter tradition, the bar beckoned, and my unaccompanied tour with 78 Squadron began.

The squadron had moved from Aden in 1967 as hostilities there ceased and transferred its twelve Wessex 2 helicopters to RAF Sharjah. As one of the seven Trucial States (the others being Dubai, Abu Dhabi, Ras al Khaimah, Ajman, Fujairah and Umm al Quwain) Sharjah was the centre of the British commitment to the area with a historic role in defending the states against any outside aggression. By 1971, the progress towards physical disengagement from the Gulf was gaining momentum and the formation of the United Arab Emirates at the end of the year, marked the termination of any RAF presence there. The size of the British contingent gradually reduced and by 1971 78 Squadron was down to four aircraft.

Tim Jenner.

The squadron's role was to support the resident army battalion based there, to provide limited support to the Trucial Oman Scouts (a British-funded paramilitary organisation), a search and rescue capability for the other squadrons based in Sharjah (Hunters, Shackletons and Andovers), and occasional support for the Sultan's army in neighbouring Oman.

It's worth reflecting on how the area has changed over the intervening fifty years. Sharjah itself was a small dusty town with very little apparent wealth although the oil and gas deposits, which were beginning to make the area so fabulously affluent, were starting to be exploited. Neighbouring Dubai was just commencing its transformation to becoming the bling-capital of the world and Abu Dhabi was trailing just a little further behind. The key indicator of progress seemed to be the construction of new roads to nowhere in particular, with roundabouts added at regular intervals to relieve the boredom. Abu Dhabi then stole a march on Dubai by adding a flyover to its programme, and the breathtaking march to the vast skyscraper-dominated landscape of today was well under way. The other four states – now all major holiday destinations – were little more than tented encampments. Apart from Sharjah, there were two other airports, one in Dubai and another in Abu Dhabi. The Sharjah and Dubai runways were parallel and only a few miles apart, resulting in some interesting, missed identifications in the usually poor visibility along the humid coast. On the odd occasion, this resulted in civilian traffic aiming for Dubai managing to land in Sharjah before a hasty and embarrassed departure.

Two Wessexes operating out of RAF Sharjah.

Landing areas in the hinterland of Sharjah were very challenging for the pilots.

The terrain was flat desert, but with the rocky mountains – the jebel – to the east rising to around 3,000 feet and running to the small outcrop of Omani territory in the far north of the Masandan Peninsula, and down into the higher mountains of Oman itself in the south. The weather ranged from the relative cool of the winter months to the extremely hot and humid days of the summer with regular temperatures in the mid-forties. Visibility was often very poor, particularly on the humid coastal strip and was sometimes exacerbated by the odd sandstorm. There was very little rainfall, although when it did occur it could result in spectacular and damaging flash flooding, especially in the jebel.

Living accommodation was in Twynham huts – a metal design which provided individual rooms for the officers' mess and with a centrally provided source of air conditioning. Inevitably this was subject to the odd failure resulting in very uncomfortable times until it was fixed. The mess itself was welcoming with good catering and a well-supported bar. One particular occasion remains in my memory. Having tottered back to bed rather late, but not the last to leave, after a jolly evening in the bar, I was a little annoyed to be awakened from my slumbers by a mate banging on my door and shouting "Come back to the bar quick – Miss World is there!" I wearily dismissed the claim, even though it was repeated with some conviction, and settled into a good night's sleep. The next morning, I discovered that an American military flight had landed unexpectedly, complete with not only Miss World (Eva Rueber-Staier), but also the legendary Bob Hope, both on their way to entertain troops in Vietnam.

The Wessex qualifies as one of the original proverbial brick-built outhouses with a very robust airframe construction that could absorb most of the ill-treatment routinely flung at it by crews and passengers alike. But its Achilles heel centred on the Gnome engines and the intake system to the two turbines. The engines were mounted in the nose pointing down towards the ground, and although a nose door was developed that positioned the air intake higher up, it was not able to filter out much debris. The sand in the desert varied tremendously in its effect on the Gnomes, but always to the detriment of the inlet guide vanes, and usually the rest of the engine.

Checks of the engines' efficiency were regularly carried out on the tie-down base. This was a crude and uncomfortable exercise which involved anchoring the aircraft to the ground, running up to maximum power, and – by varying the inputs to the computerised guide vane actuators – causing the engine to stall. By checking the point at which the stall occurred, the engine's efficiency was determined. I found it an uncomfortable exercise because no helicopter likes to have the ability to absorb vibration removed or interfered with. But, by tying the airframe firmly to the ground, the forces through the airframe at times felt severe – in effect trying to cause the earth to vibrate. The discomfort increased as power was applied, and the Wessex strained to rid itself of the constraint of the various rather Heath Robinson-looking collars and chains that were used to secure the airframe to rusty attachment points on the tie-down base. All this was a regular requirement as engines wore out very quickly – in the most extreme case resulting in two brand-new Gnomes being damaged beyond repair within twenty-five minutes total running time simply because of the ingestion of recirculating sand.

The challenge of avoiding a sand cloud was not limited to the health of the engines. The downdraft of the Wessex was sufficient to create a dense and totally opaque cloud which – in the most extreme circumstances – made hovering and landing exceptionally hazardous. The classic CFS technique in such conditions was to execute a zero-zero approach and landing in which the forward ground speed and rate of descent were both reduced to zero as the aircraft touched down. Like much of the output from CFS, this was fine in theory but not in practice, particularly in the worst desert sands. The alternative was to carry out a running landing so that the sand cloud remained behind throughout the critical phases. Surprisingly, many of the desert surfaces were flat enough to be able to use this technique safely, aided by the aforementioned robustness of the Wessex airframe and undercarriage, and not too many landings caused any significant damage to the aircraft despite often running on at speeds of thirty knots or so.

However, on one occasion, the robustness proved insufficient. The main undercarriage strut was a long near-vertical oleo fixed at the bottom to a trailing arm with the wheel attached. After a particularly energetic running landing the attachment bolt between the oleo and the trailing arm fractured. By an unfortunate quirk of geometry, the trailing arm – with wheel attached – rotated rearwards through an arc of nearly 180 degrees with the wheel coming to rest firmly wedged in the Gnome engine exhaust. The pilot was a young and relatively inexperienced first tourist, but he managed to get the aircraft back into the air, shut down the affected engine, and return to base, hoping that someone might have a good idea to resolve the situation. Needless to say this was not in any CFS textbook, or any other for that matter. After some discussion with the engineers, we concluded that it might be possible to work on the aircraft while it maintained a hover, manhandle the trailing arm back to its correct position, and put in a new bolt. Easy to say, but quite a challenge for the pilot, hovering in extreme heat at the limit of single-engine performance, for well over an hour. However, to his great credit, and aided by another who climbed on board the hovering helicopter and into the vacant left-hand seat, the operation was completed successfully. It was rewarded by a rather paltry Green Endorsement to his logbook.

Task flying in support of the resident battalion largely consisted of small-scale exercises in the jebel. Many of these involved picketing a wadi – moving platoon-size groups from high point to high point overlooking the wadi bed, securing a route through the mountains. The high points were normally no more than a jagged outcrop and therefore necessitated hovering with just one wheel in contact with the ground. Being hot and high, the performance envelope of the Wessex was limited but there was normally power in hand. However, the prospect of an engine failure was always a major consideration and planning escape routes was a primary concern. Happily, despite the well-known problems with the Gnome engine fuel computer system, there were no serious incidents.

Hot and high operations did mean that cockpit temperatures were often extremely high and the presence of a large fan on top of the helicopter did little to alleviate the problem. It was routine to carry large flasks of water to keep up fluid intake, the crewman in the main cabin downstairs would pass up regular drinks to the cockpit. A jolly jape involved filling a whisky bottle with cold tea and passing that to the pilot to the amusement, and sometimes concern, of the passengers below. The temperatures also encouraged the ingress of unwelcome insects, on one occasion my ear began to develop a severe itch – on removing my bone dome in mid-flight I discovered a large cockroach lurking in the earpiece.

There were a few large-scale exercises with the army – normally a week or two under canvas in the desert. Being away from the coastal strip meant that the heat was much drier, and, with cool nights, the camping experience was not too unpleasant. There was always an interested local presence even when apparently many miles from any habitation. On one exercise we were trialling a large rubber membrane, about twenty-five metres in diameter, to be laid on the desert floor to form a landing pad. A party from the UK came specially to set it out, taking two days to meticulously prepare the site. We awoke the following morning to find that the interested local presence had managed to squirrel it away during the night, despite it being right next to our tents. It was never seen again.

Operations with the Trucial Scouts and the Sultan of Oman's Army provided variety. A favourite task was to fly down into Oman to resupply the small outpost high in the Jebel Akhdar at Saiq. An Argosy from Bahrain would bring supplies to a desert strip at Firq which we would then transfer to Saiq – 6,000 feet up. The Argosy landing was always an entertaining diversion. As much of the resupply was water the run became known as 'Eau for Firq–Saiq' and we even had a tie created to commemorate such an inspired play on words. (Saiq is now a tourist destination with a five-star hotel reached by a motorway-standard road.) Another operation in Omani territory happened at the end of hostilities with the rebel groups in Oman when we took Special Forces to deal with the last remaining outpost in the Musandam Peninsula. To my knowledge that was the only occasion when 78 Squadron in Sharjah was involved in a shooting war – albeit on a small scale.

There wasn't much demand for night flying from the army so the majority of after-dark forays were training flights. The normal approach aid for landings away from airfields was the 'T' – five lights laid in a clear area in the shape of a letter T. The approach was made along the upright stem of the 'T' and the theory was that by maintaining the correct perspective of the lights a constant and safe angle of approach could be flown. However, without some form of natural lighting to give an idea of which way up you were, the exercise became challenging. The first time I flew to a 'T' in the middle of the desert on an overcast night with very limited visibility was memorable. Finding the 'T' was a challenge but establishing a circuit and making an approach proved more so. All that was visible was the five lights and they could have been five stars of an unknown constellation for all I knew. There were many who claimed that flying the 'T' without any external reference was impossible and Sharjah on a totally black night certainly reinforced their view.

Casualty and medical evacuations provided much of what little operational night trade existed. One particularly memorable medevac involved rescuing the victims of a horrific head-on crash between two jeeps in the middle of the Jebel. A call came through to Sharjah from the local Trucial Oman Scouts' base at Masafi. This was situated on the only road through the jebel and was in a small bowl surrounded by mountains on all sides. Needless to say, it had to be a dark moonless night and the approach to the landing site – illuminated by the headlights of two Land Rovers – was hazardous. Fortunately, the call for help came through to the bar where we were socialising with some Shackleton mates, and a plan was formulated to launch their aircraft to provide overhead cover by firing flares to illuminate the site as we made our approach. We also were informed that there were at least nine seriously injured casualties, so we had to send two helicopters. All three aircraft were launched in double-quick time and the plan worked marvellously as I spiralled down into the bowl on the first approach. But, at the *moment critique*, the Shackleton called to inform us they had run out of flares. Quick thinking by the second aircraft provided some limited lighting by firing a flare or two from the Very pistol each Wessex carried, and successful landings ensued. Happily, all casualties survived after we took them back to Dubai, including one who had some rather primitive surgery performed by the RAF doctor on the flight. The entire operation was rewarded the following day by our HQ in Bahrain complaining that the whole thing had been done without their authorisation.

One other SAR operation stood out – two super-tankers, one laden with oil and the other with gas, somehow contrived to collide in the middle of the Gulf in T-bone fashion. The squadron was tasked to do something – nobody was quite sure what – and our intrepid squadron commander rushed forth, landed on the deck of one of the tankers (still firmly wedged to the other) and offered assistance. Risk assessments had not been invented in 1971. Somehow the crews of the ships managed to sort out the situation without creating a major explosion and the squadron's aircraft and personnel remained intact.

All good things must come to an end, and so it was with 78 Squadron, RAF Sharjah, and the permanent British presence in the Gulf. Diplomatic negotiations for the ending of the Trucial States arrangement were going on throughout 1971 and, in the second half of the year, became more active as the end of year deadline approached. The senior diplomat charged with the process was Sir William Luce and I spent some time ferrying him from state to state as he discussed the plans with local leaders. Those meetings in the

Fly-past at the disbandment of 78 Squadron on 26 November 1971.

Tim Jenner escorts the British Commissioner at the disbandment parade of 78 Squadron.

smaller states were literally conducted in tents over a feast of local delicacies (sheep's eyes figured prominently). An interesting trip was to one of the small islands in the Gulf which was disputed with Iran. Strategically important, but only a few hundred yards across, and uninhabited apart from some dodgy-looking snakes, none of us wanted to leave the helicopter. The result of all the negotiations was the creation of the United Arab Emirates (UAE) in December.

RAF Sharjah gradually ran down in the second half of 1971, as did the army presence. There was a general air of 'go-home-itis' around. Much of the equipment was left behind or imaginatively disposed of. We took the squadron stock of smoke grenades and flew over a very deep canyon and then released them as quickly as possible to produce a multi-colour display worthy of the Red Arrows. There was little consideration then of environmental aspects. Visiting aircraft took the last opportunity to do a fly-past which would invariably be announced by ATC over the ops room squawk box. One warning of an imminent VC10 low pass was nothing out of the ordinary until we saw it was a BOAC flight out of Dubai loaded with passengers enjoying the view of the Sharjah runway from low altitude.

No. 78 Squadron disbanded with a very small parade and a fly-past, the silver was packed up along with the key documents such as the line book, and, by December, everyone was on their way home to the same snowbound Brize Norton that I remembered leaving eleven months previously.

CHAPTER EIGHTEEN

HUNTERS OVER SINGAPORE

TREVOR NATTRASS

JANUARY 1963 AND FOUR YOUNG pilots from the recently disbanded 14 Squadron in Germany arrived in Singapore courtesy of a thirty-four-hour flight by a Caledonian Airways Bristol Britannia. The flight was dry and staged through Istanbul and Calcutta.

We were met at Changi airport and taken the twenty or so miles to RAF Tengah. At the officers' mess, which was a large white building with porticoed verandas, colonial-shuttered windows and doors, situated on some rising ground to the west of the main runway, we were allocated rooms and moved in with the help of our Chinese batmen. Lunch was

Trevor Nattrass checks his weapon load of rockets before take-off.

followed by a quick visit to the squadron area, and then off to the resident tailor's shop, which was run by a very capable Chinese with several staff. The demonstrable work ethic of that community was immediately apparent. Measurements were taken and we ordered multiple shirts, shorts, and socks, together with bush jackets, dress uniforms and tropical mess kit. One set of the tailored shirts and shorts would be ready at the end of the day, with the rest available tomorrow, except the dress uniform and mess kit which would take three days. Moreover, it was all affordable and, as it transpired, very good quality; in stark contrast to the expensive, frequently poorly made kit available from the well-known military tailors of the day in London. Officers were required to buy their own uniform, but why did I need so much kit? It was simple, because of the heat and humidity, there was no question of using clothes for more than one day.

Next came a quick session with the doctors, where we were issued with anti-malarial and salt tablets, one each to be taken daily. We were briefed on the various bugs and diseases waiting to snare us, and the dangers of alcohol in such an environment. No one mentioned Tiger Beer, and that should have definitely featured in that briefing. Then back to the officers' mess to build mosquito net cages round our beds, to finish unpacking, shower and retire to the bar prior to meeting more of the squadron pilots. After an early night, I learned the mistake of going to sleep with a mosquito inside the bed cage, and not to have a hand against the netting. I started the next day with a very swollen right hand and some numerous

The 20 Squadron flight line at RAF Tengah, Singapore. (AHB)

bites across my knuckles and thumb, looking very pale, even moony, compared to my new squadron colleagues. It wasn't hard to tell which of us had just arrived from the UK.

New to the Far East, and ready to absorb its ethnic diversity and multiple cultures, we were unaware that the political decision for North Borneo, Sarawak and Singapore to form the Federation of Malaysia would dominate operations throughout our time in South-East Asia. The squadron had, however, already taken part in operations in Brunei to help suppress an attempted insurrection there.

Our stamping ground was essentially over the jungle-covered terrain of the Malay Peninsula, which presented a navigational challenge. There were few landmarks, and the maps were fairly inaccurate. We marked up our own maps based on our belief about what was where. It was even worse over Borneo, where the maps were overprinted with the phrase, 'largely un-surveyed'.

My early days on the squadron concentrated on air-to-ground operations and firing all four guns at both stationary and towed (splash) targets. We newcomers were introduced to one-v-one air combat, which resulted in the recognition that we knew our way around a dogfight. After one such sortie, a chastened flight commander said; "You've done that before then?"

For concentrated air-to-ground weapons training, so-called armament practice camps, the squadron normally deployed to the Royal Australian Air Force Base at Butterworth, about 250 miles north-west of Singapore, where we were able to use the air-to-ground range at Songsong on Penang Island. The detachments, typically for two weeks, had the advantage of also allowing us to explore the delights of Penang.

At base, we used Asahan Range near Malacca, which had to be manned by a detachment of squadron personnel, with the attendant need for the provision of detachment security, catering and the like. Not to mention the problems of living cheek by jowl with some aggressive local wildlife, including tigers. As detachment commander and range safety officer at Asahan, there was always the issue of one of the range party being spooked, or claiming to be spooked, by a wild animal and loosing off a few rounds from an available weapon, usually a Stirling sub-machine gun. The downside of such an act was the nightmare of accounting for spent ammunition. During my stint as detachment commander, such liberties were not permitted, as I had no desire to have a pissed-off tiger roaming round the campsite.

From time to time, the squadron was required to supplement the headquarters staff in the Combined British Forces Borneo HQ (COMBRITBOR) in Brunei and I had the pleasure of being dispatched there to be an operations officer in the Air Transport Operations Centre (ATOC) for a month. The staff were accommodated in dormitories in the closed local girls' school, with tiny beds even for someone only five foot six inches tall. However, during my off-duty hours I managed to fly as a co-pilot in Army Air Corps Beaver aircraft and visited several jungle landing strips associated with small army forts. The 400-foot airstrips were quite a contrast to the 9,000-foot strips I was used to. Not only was I treated to that novel experience but also, for the first time, met loin-cloth-clad tribesmen, complete with blowpipes and poisoned darts, all of whom I towered over, which was something of a first for me. They were delighted to receive twenty Rothmans cigarettes in return for showing off their hunting kit.

In 1963 the wars in Indochina had not yet gained their later prominence, but in late May the squadron was deployed, with elements of 45 Squadron, to Chiang Mai in northern Thailand to demonstrate the UK commitment to the South East Asia Treaty Organisation (SEATO) as a deterrent to the warring factions in Laos. The airfield, which had little permanent infrastructure, was subject to monsoon flooding, which meant that the tented accommodation needed bunds around each tent to keep the worst of the water at bay. Two layers of duck boards kept our beds reasonably dry but seeing small fish swimming freely in the water in our tents soon ceased to be a novelty. Nevertheless, and despite the lack of facilities, we were British officers and thus meals taken in the tented officers' mess required long-sleeved shirt, collar and tie, slacks and shoes, so-called planters' dress. Thus, before dinner, one changed, rolled-up the trouser legs, and with shoes and socks in hand, 'plodged' through the water to the mess, where towels were on hand to allow the restoration of officer appearance.

The squadron was supported by a seconded, army ground liaison officer (GLO). His job was, essentially, to explain the niceties of any ground war to enable us to understand where our own forces were. Captain Jimmy James was a guard's officer of suitable stature. He was accompanied by a driver (Corporal Clayton) who couldn't drive but could type, and a typist who couldn't type but could drive. At the end of each day Clayton would collect Jimmy's boots and take them to the other ranks lines to be dried and cleaned for the morrow. One of the abiding sounds of the detachment was Jimmy's stentorian "Corporal Clayton!" each morning, which resulted in a responding call of "Sir!" and the corporal would then appear with newly shined boots in hand. Jimmy would inspect them, put them on and immediately step over his tent bund into the water...such fantastic routine and discipline kept us sane, or at least amused.

From a social standpoint, the local town of Chiang Mai offered little in the way of entertainment, with the odd very poorly lit nightclub and the inevitable bar girls. Beer was the drink of preference, with Thai whisky, called Mekong passable, and very potent, even when mixed with 7 Up or Coca Cola. Transport for all was usually by three-wheeled trishaws peddled by the 'driver'. So, trips to and from town were sometimes enlivened by races arranged between the drivers with us cheering our respective driver on from the seats behind him. The prize for winning the race was the usual fare, plus a few extra baht. Fun for us, but I doubt we ever considered that we were exploiting the poverty of the drivers. It was a hard way for them to earn what was effectively a tip, although my recollection is they seemed to enjoy the break from their more mundane tasks.

On 17 June, I was part of a formation tasked to escort the King of Thailand and his entourage in the royal aircraft back to Bangkok and away from their summer home in the hills near Chiang Mai, as there was a risk from one or more of the various factions in Laos. For the guys concerned this was a bit of a non-event as there was no known air threat, but the impression created was important as we duly 'protected' the two DC-3 Dakota aircraft on their route back to Bangkok.

In August 1963, I was detached to fly with 28 Squadron in Hong Kong. This was a regular detachment which allowed 28 Squadron pilots to come to Singapore for simulator training, and to operate in much less restricted airspace than that round the Hong Kong colony, limited by the proximity of Communist China.

Under the guidance of Ian 'Chunky' I'Anson, I did two familiarisation trips to learn the colony boundaries, and quickly discovered the joys of virtually unrestricted low flying and the licence to practise attacks on virtually anything which took my fancy.

A Hunter on finals to Kai Tak Airport, Hong Kong.

We operated from the airfield at Kai Tak, with a runway which stuck out like a finger into the harbour. I took off in Hunter FGA.9 (XJ622) having been briefed to practise simulated gun and rocket attacks against moving targets, specifically the HK to Shanghai express, which was due to leave Kowloon Station shortly. I slipped up into Tolo harbour then came back over the peninsula, and picked up the train, adopting a look of grim determination, shades of rocket-firing Typhoons in World War Two. All this action was recorded on my gun-sight camera. After several attacks, the train pulled into a station (big clue here as the only station was at the border crossing) however, as the train no longer presented a decent target, I left the scene of a crime to which I was oblivious, and after a quick run around the harbour I returned to Kai Tak to be greeted, not just by the ground crew, but also the OC 28 Squadron and others.

"Where have you been?" Still the realisation hadn't dawned. Cinefilm confiscated, and reports from the border post of a Hunter overflight. I was the only one airborne. Great attacks on a moving train on the gun-sight film, and great pictures of the border station at Lo Wu. Not only had I infringed the 'never to be crossed' Outer Flying Boundary, but I had also been into Red China. Marched in to see the station commander; formal questioning under oath, with an escort officer (Flying Officer Peter Headley) and placed under close arrest pending a Board of Inquiry. My day was not working out as expected.

There wasn't much to enquire about that wasn't already known, but the inquiry visited the border post at Lo Wu, where the eyewitness evidence confirmed my worst fears. I also learned that the Chinese air force (correctly known as the Peoples' Liberation Army Air Force [PLAAF]) had scrambled their battle flight to counter the intruder. On reflection, it was clear that running into a pair of MiG-17 fighters could have made my bad day much worse. In truth, my mistake deserved little more than a formal apology to the Chinese, however, as the UK did not have diplomatic relations with the communist government our diplomats were not best pleased with the RAF and me.

More days in close arrest followed and then, after a further formal meeting with the station commander, my case was remanded for summary disposal by the commander-in-chief, one Denis Crowley-Milling of Battle of Britain fame. Pending that interview, I was released into open arrest, which effectively meant I could not return to work but could roam, although not leave, Hong Kong. A 22-year-old on holiday (effectively) in HK courtesy of the RAF; I got up to no mischief.

The day of the summary disposal of the charges against me arrived. I was facing the prospect of everything from a court martial and dismissal from the service to a slap on the wrist. So, in best uniform, and once again under close arrest, I set off for Air Headquarters on Victoria Island with my escort on the RAF launch used for transit between Kai Tak and the island. The whole thing was redolent of being flogged round the fleet, or perhaps being keel hauled, and now I was going to be late, because halfway across the harbour the launch engine stopped. We drifted on the current, but after several tense minutes the coxswain stripped off and, knife clenched in teeth, dived over the side to hack off the rope which had fouled the propellers. Feeling I was about to be the sacrificial lamb on the altar of the Foreign Office, it did not seem that being late was going to help my case. Exotic Hong Kong had become a designer nightmare.

A tense wait in the C-in-C's outer office; marched in to meet my fate, breath held for most of the time. The C-in-C announced that he was awarding me a reprimand, the lightest punishment available. How I avoided smiling is beyond me. In his outer office I was just starting to breathe again when Sir Denis appeared and said something to the effect that the Foreign Office was unhappy with me, as a grovel to the Chinese had been necessary. He then smiled and said, "If this had been wartime you would probably have won a medal, let me buy you a beer and, by the way, you are on the transport out of here this evening and are banned from Hong Kong for life." Back in Singapore, my then boss viewed my error as a major affront to him and I was in bad odour for several weeks.

Many years later, at an air show at Shenzhen Airport (from which the MiG-17s launched against my incursion) I met with the head of the PLAAF. Through the interpreter, I recounted

Facilities at Chiang Mai, Thailand were basic. (AHB)

this story to his evident amusement, probably about the diplomatic embarrassment rather than my trip to see the C-in-C.

Back in the Singapore routine, the tension with Indonesia continued to grow, and the base continued to provide a home for the air wings of one or more of the RN aircraft carriers. The Royal Navy had, from time to time, three carriers in the South China Sea during the period, when it was not unusual for Tengah to host some of the embarked squadrons, while the carrier docked for some routine maintenance. Moreover, as the year turned, I was part of a four-ship which was tasked with escorting the Malaysian president around Borneo, which was followed shortly thereafter by the establishment of the Malaysian Air Defence Identification Zone, which was policed by detachments of Hunters and Javelin aircraft based in Kuching and Labuan off the coast of Borneo. Operating from primitive, largely tent-based facilities, we quickly formed the unofficially titled Borneo Jet Force, with suitably irreverent crest.

Another detachment to Thailand, this time to Udon Thani on Exercise Air Boon Choo. The airfield was also home to an 'airline' called Air America, which was equipped with an eclectic array of aircraft such as the Scottish Aviation Twin and Single pioneer aircraft, which were ideally suited to short field operations. Here was evidence that things north of the Vientiane River (the border with Laos/Indochina) were becoming serious as this was the 'airline's' favoured operating area. The aircraft would disappear at first light, and it was

A Hunter FGA.9 of 20 Squadron low over the Borneo coast.

not unusual for some of them to come home with battle damage. Our detachment was uneventful, spent mostly doing forward air control with US ground forces. However, the war to the north was brought more sharply into focus when, during a visit to Bangkok, we saw a USAF F-101 Voodoo aircraft return to base with a large hole in it.

As we approached mid-August, and the threat of insurgent activity by the Indonesians increased, all military leave (including honeymoons) was restricted to Singapore Island. Not a hardship you might think, but in 1964 Singapore was not the high-rise metropolis it is now, indeed there was only one air-conditioned building (called Cold Storage) and the tallest building was a hotel with six floors. However, my bride and I were fortunate that some friends of ours were not affected by the travel ban and they loaned us their house and servants, which is where we repaired after the nuptials. Not exactly a secret location, and we were soon joined by a large group of squadron members. Realising that going with the flow was the only likely route to some later privacy, we all went off to the cinema at midnight to see the Far East premier of 'A Hard Day's Night'. Romantic stuff, especially the decision of the 'team' to join us afterwards. Perhaps not quite what we had intended.

We settled into a lazy routine, but on the Thursday morning, 27 August, the phone rang at about 0730 and my flight commander, 'Black' Ferguson (actually Donald Currie Ferguson) said: "Trevor, this is not a honeymoon wind up, pack your things and both of you come into the station as quickly as possible." So, we did and by mid-morning I was on 'Battle Flight' duties on the aircraft readiness platform underneath the wing of my aircraft. President Sukarno was ramping up the anti-Malaysia rhetoric, but the immediate worry was (as ever) one of the RN's carriers transiting the Sunda Strait, which seemed to offer an excellent opportunity for the Indonesians to score a propaganda coup. We were armed ready to help our dark blue colleagues, but nothing happened. Instead, we started morning and evening combat air patrols against possible Indonesian insurgents approaching Singapore by sea. So, just three weeks after our nuptials I flew my first 'in anger' sortie, firing cannon and sixteen rockets against some Indonesian parachutist near Malacca on the Malaysian peninsula. Then, same again the following day against insurgents who had landed an Eye Island under the approach to Changi airfield and who were making a nuisance of themselves. These guys watched us start our attacks, as they clearly felt 'safe' in and around the World War Two bunkers and the like, but they quickly surrendered to the deployed Gurkhas after we finished. Bit of a rock and a hard place really.

Our operations were not without cost and in quick succession we lost Flying Officer Doug Clavering during a dawn patrol over a misty Straits of Singapore without satisfactory explanation, although disorientation probably played a part. Then, the vastly experienced Flight Lieutenant Bob Shields, who ejected following a hydraulic fire after take-off from Labuan. He landed close to the airfield, as he tried to recover the aircraft. Pete Riley and I

Armourers at work on the gun pack of a 20 Squadron Hunter. (AHB)

went with the rescue chopper to pick him up; he was conscious but in a bad way, and he died soon after we got him into sick quarters. As the Doc cut Bob's flying suit open to assess his injuries, he just looked to the heavens and shook his head. There was nothing to be said, it was a very sobering moment. Bob said, "Thank God for Martin-Baker" (the ejection seat manufacturer), but in fact the seat separation system had malfunctioned thus preventing full deployment. They were his last words. We understood accidents were an inevitable part of what we did when the challenges of fast-jet flying met human or technical frailties.

Aircraft losses continued when Pete Martin had to eject following an engine flame-out during an aerobatic practice over the airfield at Tengah, and then Heinz Frick had to eject following loss of control during a dogfight. This latter event was reported in the main daily newspaper *The Straits Times* as 'Pilot ejaculates and lands in a swamp', a front page that quickly found its way into the squadron diary and line book.

The frequent roulements to Kuching and Labuan continued for border patrol duties, including flying escort for HRH Prince Phillip during a visit to the region. The patrol sorties were largely uneventful, but the confrontation on the ground was very active. On one such patrol when my number two had aborted, I was following the well-trodden route back towards Kuching, visiting various army jungle forts along the way. The routine was to give each fort a call on the radio and then give them a fly-past or two, as a morale booster. However, on this sortie, when I called the fort at Siri, which was only about fifteen minutes flying time from Kuching, all was not well; sounds of gunfire on the radio, and an agitated army officer explaining they were under attack by about 200 insurgents. He said he would mark their positions with smoke and could I please put down some suppressing fire. I could see the fight and the smoke, but the rules of engagement required me to obtain clearance to fire at the bad guys. I did a couple of high-speed passes over the insurgents, while waiting for clearance, which required me to keep zooming up to about 10,000 feet to talk to Kuching. Clearance, however, was not forthcoming, and low on fuel I was forced to return to Kuching, abandoning the thirty-four troops in the fort to their own devices; my request for a further sortie to give them some air support was not authorised. Retrospectively, and seeing the casualties brought into Kuching by helicopter later in the day, I should have ignored those rules and I promised myself I would do so in the future. Apparently, there was no desire at the diplomatic level for the UK to be seen to be using jets against insurgents; which begs the question, why were we there? A perfect example of 'unlimited liability'.

On another occasion, to the east of Kuching, my wingman and I came across a C-130 Hercules, only operated by the Indonesians, or so we were told. This aircraft was clearly violating Malaysian territory; we closed up ready to make it land one way or another only to see Royal Australian Air Force roundels on the wings and fuselage. Our various briefings had neglected to mention that the Aussies were going to appear. Never a dull moment and that oversight could have seriously spoilt someone's day.

My stint in Kuching came to an end and without warning, although not unexpected, on return to Tengah I received a posting notice to 229 OCU at RAF Chivenor in Devon to be an instructor on the Hunter Operational Conversion Unit. So, in a routine with which we were to become very familiar, Liz and I packed our possessions, and via our first-ever long-distance flight in a jet transport, an RAF Comet 4, we returned to the UK.

CHAPTER NINETEEN

THE CANBERRA B.15 IN FEAF

JEFF JEFFORD

As a first tourist navigator, I joined 45 Squadron at Tengah in the summer of 1962. At the time we were the last squadron still flying, by now rather tired, Canberra B.2s and were about to be remounted on B.15s. Re-equipping the squadron was done on a self-help basis with crews flying a B.2 back to the UK and returning in a B.15, the first of which arrived in September.

Before the ferry programme was complete a crisis occurred in-theatre when, beginning on 8 December, there was an attempted revolution in the Sultanate of Brunei. Although the Far East Air Force (FEAF) immediately responded to the crisis, 45 Squadron was unable to contribute substantially because its B.15s were not yet operational. That said, it is doubtful that FEAF ever seriously considered using its Canberras in an offensive role as the last two B.2s, nominally the only calibrated bombers 'east of Suez', were allowed to depart for the UK

Jeff Jefford (right) and his Canberra crew of 45 Squadron.

two days later. Although notionally capable of being used, these two aeroplanes, WH667 and WJ986, were the tattiest of the old fleet. In fact, several bits of WJ986 had flown home already as they had been removed to make other aeroplanes fit for the trip. It had been the 'hangar queen' for some time and the squadron's engineering officer had been hoping that it would only be necessary to patch it up sufficiently for it to make the hop to Seletar where the local maintenance unit could have quietly put it out to grass. 'Their Airships' decided otherwise, however, and a great deal of effort had to be expended on preparing WJ986 for the long trip back to the UK.

I was a participant in the last of the B.2–B.15 shuffles, and we were soon to learn that the attempt to patch up '986 had been only partially successful. Flying as a pair and taking it in turns to lead on alternate legs, we were No. 2 to '667, on the Calcutta–Karachi leg when one engine suddenly stopped, precipitating the aeroplane into a steep spiral dive, accompanied by the broadcasting of a Mayday message. Several thousand feet lower down the driver, Flight Lieutenant Stuart Pearce, announced that he had the situation under control again and cancelled the emergency. But it took all his persuasive skills and the passage of a respectable period before he was able to convince his navigators, Flight Lieutenant Phil Mason and this writer, that normal service really had been resumed. There followed some embarrassing confusion as to which of Karachi's several airfields was supposed to be our destination. Why? Because, as No. 2, we hadn't paid quite as much attention as we should have done pre-flight – as 'Wing Commander Spry' used to say, in practically every edition of the RAF's contemporary flight safety magazine, *Air Clues,* 'I learned about flying from that'. In the meantime, our intrepid leader had flown serenely on, blissfully unaware that he had lost his companion, and the limitations of the B.2's single VHF radio meant that we never re-established contact with him. Fortunately, we picked the right airfield – Drigh Road – and landed on one of its parallel runways as No. 1 landed on the other. The aeroplane proved to have an unserviceable fuel pump which took two days to rectify. Thereafter the trip went as planned until the last leg, from Malta to the UK. Shortly after take-off, '986 lost its only radio and, having used frantic hand signals to explain the situation, Flight Lieutenant Rupert Butler led the transit across France and into Lyneham with WH667 doing the honours for both aeroplanes.

While steady progress was made adjusting to the new aeroplanes and their significantly expanded capabilities, compared to the old B.2s, the familiarisation process was hampered by the constant detachment of individual aircrew because of the ongoing tension in Borneo which had since spread to Sarawak and Sabah. As a result, the squadron usually had four men away, filling one-month unestablished operations officer slots with HQ FARELF at Phoenix Park in Singapore and/or at Labuan, Brunei and Kuching. Interesting as these jobs were for junior officers, the detachments prolonged the process of consolidation with the new aeroplanes.

Canberra WJ766 drops its load of 1,000-lb bombs from the bomb bay.

Another, constant diversion of effort is worthy of mention. There were no range safety officers established for the three air weapons ranges in Malaya and they were manned by the squadrons which used them. It was usual for the Australians at Butterworth to look after their local range at Pulau Songsong, an island to the north of Penang. China Rock, off the south-east tip on the mainland, became 45 Squadron's regular commitment. The routine was to leave, accompanied by two or three airmen assistants, from Seletar in an RAF launch early on a Monday morning. A couple of hours later the party was deposited on the beach with a week's supply of rations and the boat departed, leaving the team marooned until teatime on Friday. The RSO would already have a copy of the week's range bookings (or 'dream sheet') and the first tasks, made easy by a well-thumbed check list, were to fire up the pre-positioned Land Rover and have both quadrant huts manned and the radios working in time for the first scheduled bombing detail. The third range, Asahan, which was on the mainland some seventy miles (about three hours') drive north of Singapore, was not permanently active and when it was, the chore of providing the RSO generally fell to 20 Squadron, although the odd 45 Squadron man could be seen there too from time to time.

There was a significant diversion in May 1963 when the USAF's 3rd Bomb Wing sent a detachment of its B-57s down to Tengah on an occasional mutual exchange programme – Exercise Joss Stick. Opportunities were provided for the navigators to swap seats and it was an eye-opener to fly in the B-57. The cockpit was roomy, comfortable, and provided tandem seating (with built-in ashtrays) under a clear canopy, permitting a panoramic view of the outside world – in contrast for the navigators to see out of the RAF's aeroplanes they had either to perch on a fold-down occasional seat next to the pilot or lie in the bomb aimer's position wearing a neck-cricking bone dome; in either case this involved abandoning the, albeit limited, security offered by their ejection seats. Even better, the B-57 could be taxied with the lid open and when it was closed prior to take-off the air conditioning was instantly effective and the cockpit momentarily filled with fog as the moisture was rapidly condensed out of the air. The RAF navigators were all eager for a second trip but the Americans, when they emerged, sweat-soaked and gasping for a cheroot, from the dark cubbyhole which provided the working environment for their British counterparts, seemed curiously unenthusiastic about repeating the experience.

In the early 1960s the situation in Indochina was slowly deteriorating and in June 1963 SEATO decided to exert itself by mounting a major exercise to remind everyone that it was there. For Exercise Dhanarajata, which was to be held in Thailand, 20 Squadron's Hunters were detached to Chiang Mai, for which deployment 45 Squadron provided a weather-recce aircraft to lead the way. During the exercise, the Canberras either turned round at Don Mueang to fly another sortie on the way home or landed at Chiang Mai to spend the night before flying a lo-hi sortie back to Tengah the following day. All such transits involved transmitting a position report shortly after entering/leaving Thai airspace which invariably sparked intensive discussion as to the correct pronunciation of the name of the designated reporting point – Phuket. It is unlikely that many crews got it right.

The termination of Exercise Dhanarajata on 24 June was marked by a ceremonial fly-past over Bangkok; this had all the makings of a potential disaster. It was to include aeroplanes of every type and every nation was involved. This covered everything from RNZAF Bristol Freighters to USAF RF-101s. At the briefing it quickly became apparent that the Thai group captain in charge was having difficulty working in English and Wing Commander G. W. Johnson, Tengah's OC Operations Wing and leader of the Commonwealth contingent, offered to help out. It was a masterly display of impromptu hot planning. The wing commander first divided the assembly into those with propellers and those without, then sub-divided them again into groups which might reasonably be expected to be able to keep up with each other. Each group had then to agree on the speed at which it would make its fly-past and select a suitable IP. Johnson decided the order of the procession, slowest first, fastest last, and gave each formation a height to fly and a time to be over the reviewing stand. It was rough and ready but, if not entirely smoothly, it worked.

The jets were to assemble in three groups, Thais, Commonwealth and USAF, then join up in that order before making their final run over the city together. The selected IP was a pagoda and the Commonwealth formation, nine Canberras, three-each British, Australian and New Zealand, eight RAAF Sabres and eight Hunters, arrived first, as had been agreed, and set up an orbit beneath

Aircraft participating in Exercise Joss Stick.

scattered cloud at about 2,000 feet. As the time to leave the IP approached there was still no sign of the Thais, who were supposed to be leading. A call on the radio revealed that the sixteen F-86s were there too, but above cloud. Their leader cheerfully announced that they were coming down, immediately. Johnson promptly led his twenty-five aeroplanes out of the way to avoid the imminent Kamikaze attack. By the time that he had recovered and tacked on behind the Thai Sabres the formation was late and it set course for the parade at, for the Canberras, a rivet-popping 400 knots, instead of the 300 knots that had been negotiated. The American component arrived at the IP slightly late and had to use afterburner to catch up.

As always (well usually) happens, the gaggle sorted itself out for the final run across the reviewing stand as the last of the heavy prop jobs cleared the area. But there was one more surprise in store. As the formation thundered up Bangkok High Street, the air was full of smoke trails and fluttering leaflets which were being liberally distributed by a flock of Thai army L-19s – of which not a word had been said at the briefing! Fortunately, the pilots of the Bird Dogs appeared to be aware of what was going on and they stayed below the fly-past aircraft. Once across the city the formations broke up into their constituent sections and dispersed to their various operating bases under private arrangements. No. 45 Squadron's three aeroplanes were safely on the ground at Don Mueang in time to see the RAF's heavy transport element, a Beverley and two Hastings, which had entered enthusiastically into the spirit of the thing, re-join the circuit in echelon for a lumber in and break.

In the months leading up to the creation of the state of Malaysia on 16 September 1963, President Sukarno of Indonesia had begun to promote what came to be known as 'Confrontation', a state of armed political and economic hostility, sometimes including direct military involvement, but stopping short of outright war. From mid-1963, bands of guerrillas had begun to make incursions across the border from Kalimantan and the frequency of these increased in the autumn. It was not known how far the Indonesians would go but the possibility of a direct attack could not be ruled out. While the army established forward posts in Borneo and began patrolling the border, the British government began to reinforce its forces in the Far East.

As part of this overall pattern, it was decided to investigate the possibility of using Kuantan as a dispersal airfield. Plans for this had been in the mill for some time and from as early as August crews had been carrying out the occasional roller landing there. The squadron mounted Exercise Merrydown, a full-scale deployment to Kuantan with all eight of its B.15s, between 2 and 8 October. Beyond a runway, some frying-pan dispersals, a bulk fuel installation and an ATC shack, Kuantan boasted few other facilities and those that did exist were off-limits. The squadron pitched its tents on the airfield, dug its latrines, set up its field kitchen and operated in much the same way as its predecessors had done in the Western Desert in 1941 – such is progress.

There was an interesting exercise in March 1964 which involved dropping live 1,000 pounders on a range at Batti Malv in the Nicobar Islands. On one of these sorties the CO's aircraft developed both a fuel leak and an oxygen leak. Initial hurried calculations indicated that the aeroplane would have insufficient oxygen to get back at high level and insufficient fuel to make it at low level. Squadron Leader Brian Carruthers wisely decided that he did not need two tons of extra weight and directed Flying Officer Burr to get rid of the four bombs they were carrying. The switches having been set to ensure a safe release, the bombs were jettisoned into the sea from 40,000 feet. Bob Burr continued to watch after they had disappeared from sight and eventually observed four bursts of fire as the bombs detonated on the surface of the sea. After reassessing the situation, the crew concluded that they could safely make it back to Tengah after all and an armament officer was requested to meet the aircraft. From the position of the cockpit switches and the absence of the safety lanyards in the bomb bay the latter was able to confirm that the bombs had been jettisoned 'safe'. When informed that they had in fact gone off, he pondered for a moment then opined that, since many of the bombs were quite old stocks which had been cooking in the bomb dump since Lincoln days, it was possible that they were so unstable that they might well explode under extreme conditions. Not overimpressed, the CO asked what would happen if the bombs were dropped via the pilot's emergency jettison facility in the event of an

No. 45 Squadron dispersal at Kuantan, Malaysia during Exercise Merrydown.

engine failure on take-off. The armourer thought it 'unlikely' that they would explode under those circumstances – but he would not recommend it.

Still determined to demonstrate its solidarity and resolve, SEATO sponsored Exercise Air Boon Choo which ran for the whole of the last week of April. No. 45 Squadron participated daily, the aeroplanes turning round at either Don Mueang or Udon Thani. Flying an exercise sortie out of the former could be an interesting experience. Don Mueang was a very busy airfield, which served as both a Thai air force base and Bangkok's international airport. It had two parallel runways, both of which seemed to be under the control of one ATC officer who, understandably, sometimes lost count of whether an aeroplane was number sixteen or seventeen to land. Once on the ground, visiting military aircraft were given no special treatment and were simply parked facing the main passenger terminal in vacant slots on the bustling main ramp between DC-8s, Boeing 707s and other commercial types. Two squadrons of Thai F-86s lived at one end of this stretch of concrete and from time to time half a dozen of these would taxi at high speed to the end of the runway, scattering the neat columns of fare-paying passengers as they were being shepherded to or from their aeroplanes.

After a hurried mission brief under the wing, getting airborne from Don Mueang on an exercise sortie could be quite exciting. Reaching the end of the runway was done by emulating the Sabres and a request for permission to enter the active runway was usually answered with an immediate take-off clearance. Flying Officer Paddy Langdown received one of these automatic responses on one occasion but took the precaution of looking before he moved forward, which was just as well as the runway happened to be occupied by a pair of F-102s which had just been scrambled from the opposite end.

SEATO sponsored another major exercise in early June 1964, this time in the Philippines. Exercise Ligtas was essentially an amphibious assault on the island of Mindoro mounted from a fleet of seventy-five ships which included two British and two American aircraft carriers. When the main assault went in it resembled, from the air, a re-enactment of D-Day in 1944. In the wake of the recent Air Boon Choo, SEATO had called for only token air participation in Ligtas, but even so this ran to some 300 aircraft. FEAF needed to be seen as players, of course, and among its participating units was 45 Squadron which detached four aircraft to Clark Field. As the formation approached Clark the leader called up and requested advice as to where he should park his aeroplanes – after all he had a whole half squadron here. He was advised to put them at the end of the B-57 line. This turned out to be about sixty aeroplanes long, which rather put things in perspective.

Meanwhile, relations with Indonesia had deteriorated and in mid-September a general state of emergency was declared, and 45 Squadron was ordered to exercise its dispersal option. On the 17th all eight of the squadron's bombers, half of them loaded with 1,000 pounders and the others with rocket projectiles, their crews packing sidearms, deployed to Kuantan. The advance party had already erected the tents but had not yet had time to dig trenches, so this task became the first order of priority. A somewhat unenthusiastic start on this chore had already been made when a group of Gurkha soldiers was spotted in a remote corner of the airfield. We had no idea what they were doing but, whatever it was, it involved a trench-digging machine. Putting on his best public-school accent, the better to communicate with the British officer in charge, Flying Officer Pete Sykes went off and returned shortly afterwards with the machine and its operator and our trenches were dug in no time. Several of them contained what appeared to be shards of pottery, although little significance was attached to these archaeological curiosities at first. They turned out to be the shattered remains of the pipes which drew off and dispersed storm water from the main monsoon drains running alongside the runway. This became only too apparent after the first heavy rains as the trenches promptly filled up with water.

Like Tengah, Kuantan had also sprouted some 20-mm Oerlikons, drawn from naval stocks and now mounted on concrete plinths in the vicinity of the camp site. One of the navigators, Flight Lieutenant Don Gibbons, appeared to know how these things worked and, shortly after the arrival of the air echelon, he checked out all the aircrew on the complexities of the harness and the cocking and firing procedures. That evening the CO briefed the squadron on the likelihood of an Indonesian dawn strike and explained that if this happened the duty crew would sound the alarm by banging on a steel triangle hanging outside the ops tent. Whoever reached them first were to man the guns; everyone else was to take cover in the trenches. At dawn the next morning the alarm sounded, and everyone dived into a trench. The CO, Squadron Leader Brian Carruthers, exhorted his troops to man the guns but was confronted only by rows of heads sticking out of the ground like so many vegetables. Once they had been reassured that this was only a drill there was no shortage of volunteers willing to pose out in the open with a gun, like so many pale imitations of Burt Lancaster in From Here to Eternity, but it had been an illuminating practical demonstration of the instinct for self-preservation at work.

When initially deployed to Kuantan, the squadron was held at a high state of readiness from dawn to dusk and no flying was permitted. During this interlude an intelligence officer joined the squadron from the UK. He had been one of two qualified linguists who had been ordered to report to Lyneham with their tropical kit and prepared for an indefinite overseas detachment at an unspecified location. As their Britannia staged its way eastwards, they had enquired at each stop whether anyone wanted them. One was expected at Khormaksar where he was to assist in the Radfan campaign; the other was awaited at Changi. He was promptly given a pistol and put on the daily resupply Hastings which delivered him to Kuantan. It had been an impressive demonstration of the RAF's stand-by

The wreckage trail of Jeff Jefford's crashed Canberra.

system in operation. The only problem was that the chap who had stayed in Aden spoke Malay while the one at Kuantan spoke Arabic. Close!

After a while, the CO began nagging HQ 224 Group for permission to do at least some flying from Kuantan and authority was eventually granted to have one aeroplane airborne at any given time. The squadron had recently acquired a single B.16 which differed from the B.15 in being fitted with a sideways-looking radar, Blue Shadow. Its installation meant that there was only one ejection seat in the rear of the cabin and one of the navigators had to make do with wearing a harness to which, in the event of a crisis, he would clip a chest parachute pack before exiting via the cabin door. When this aeroplane, WT370, appeared on the flying programme on 23 September the parachute turned out to be overdue for servicing. There was a brief inter-nav squabble which this writer won, and he flew the trip. After a short low-level navex we returned to Kuantan for some circuit work. On the first overshoot the starboard engine failed to respond and by the time that the pilot had ascertained that it was not simply a surge, to which the Canberra's Avons could be prone, the aeroplane was too low, too slow, and no longer aligned with the runway. Avoiding the temptation to increase power on the good engine, which would only have rolled the aeroplane onto its back, Flying Officer Sykes kept it airborne as long as he could before, having finally run out of airspeed, altitude and ideas, bellyflopping, fully stalled, but wings level, into a rubber plantation. Like most Canberras at that time, the aircraft was still fitted with Mk 1 ejector seats, which were rated at 1,000 ft and, ideally, 250 kts, so ejection had never been an option. The aeroplane broke up and burst into flames but both occupants were fortunate to escape with back injuries from which both subsequently recovered. Recovery took several months however, and this turned out to be my last trip in a Canberra, which brings this chapter to a close.

CHAPTER TWENTY

SUPPLYING FAR EAST OUTPOSTS

COLIN CUMMINGS

HAVING JOINED THE RAF IN late 1963, eighteen months later I was into my first tour as an equipment officer and was based at Tern Hill, looking forward to my first summer ball. Out of the blue came a REFORS detachment to Kuantan, an airfield on the east coast of Malaya.

I made the cardinal sin of buying some KD uniform, rather than waiting until I was in theatre and having bade my girlfriend a fond farewell (she said she would wait until hell froze but I have to admit it was subsequently, a cold winter) I set off via a British Eagle flight from Heathrow and via a couple of refuelling stops was deposited in Singapore and the famous Changi Creek Transit Hotel.

Donning my new KD, which along with my white knees identified me as a 'moonie' newcomer, I presented myself to HQFEAF. The welcome I received might best be described as 'cool' and I was sent to Changi village to procure some acceptable local KD, before going to Tengah for a briefing on my job. The visit to Tengah mirrored the reception I had received at HQFEAF, and I was

Colin Cummings at RAF Labuan.

The domestic area at Kuantan.

glad to be whisked away for an hour's flight to Kuantan. On arrival at what was a bare base – the only permanent building was the ATC and civil aircraft terminal, although six huge fuel tanks, soon to be my responsibility, were part buried nearby – my predecessor was keen to be off to civilisation and, as used to be said of *News of the World* reporters, 'he made an excuse and left'. Kuantan had been re-activated during the Indonesian Confrontation period and Canberra squadrons were deployed, in rotation, now from Germany and Cyprus for three-month detachments.

My new staff occupied a large canvas marquee and in total we were five, with one man living an almost nomadic existence at the bulk fuel installation. The domestic site had several other marquees which acted as messes and the officers had two, one laughingly called the dining room and the other, equally ridiculous, as an anteroom. The accommodation for officers was in tents arranged in two parallel lines from the 'mess' towards the showers and on to the Elsan loos. The showers had been built by 5001 Squadron of the Airfield Construction Branch and comprised a hardstanding with soakaway drains, and whilst cold water was piped to wash basins, hot (all things are relative) water was drawn off a 45-gallon drum at the top of a tower. One learned quickly that the sun provided the heat in the water and hence showering at lunch or late afternoon was advisable. The base was in a malaria area, and I had started a course of tablets before leaving the UK.

I was allocated a tent at the end of one of the lines and the floor was raised with a duckboard as it rained every afternoon at 16:00 and anything on the ground became soaked, probably smelled dreadful as it dried out, and rotted rapidly. The bed itself was a camp bed, but I quickly ensured it was replaced by a typical iron bedstead, with a mosquito net hung from the tent's ridge. I spent my first night in complete dread and was convinced that every snake in Malaya was making its way to where I was camping, possibly along with the occasional tiger strolling in from the central hills. My fears were not allayed when one of the air traffic control officers said that he had prodded a snake that was on the runway, which he was inspecting, and had it rear up at him. For the remainder of my time, I had a torch to hand and a stick if I needed the little boys' room at night.

I quickly tried to get some flying and OC 249 Squadron arranged for me to fly on a four ship of Canberras which were to 'attack' Tengah. The sortie went extremely well for me and on the return, we passed over the wrecks of HMS *Repulse* and *Prince of Wales,* lying in the water off Kuantan and visible at certain states of the tide.

I was told that my job, in the case of an attack (presumably by the Indonesian air force) was to oversee a 20-mm cannon, six of which were mounted near the runway. Training amounted to some basic reports to 'control' using the field telephone. The guns were mounted on firm bases and surrounded by a sandbagged emplacement, and it was advisable to keep your head below the lip of the emplacement or risk being hit by the traversing barrel. This all seemed amateurish to me but worse was to come. An exercise of the arrangements was planned and as the siren went off, everything was dropped and those involved sprinted to their posts. I reported my team 'closed up' and was told to standby. The truck carrying the ammunition sped past my gun and when I complained, I was informed that there were only five pans of ammunition, and it was my gun's turn to go without.

The Canberra squadrons from Cyprus rotated through Kuantan and 249 gave way to 6 Squadron and there was also a change of ground crew including admin staff and one of the new crop was an SAC Johnson, posted to 6 Squadron direct from the UK to the detachment in Malaya. When he reached Kuantan, he was asked his trade and stated he was a steward and was promptly allocated as the officers' mess barman, a role in which he was very good.

After several weeks he went on a walk in an area near a kampong (village) where there were several tracks. Some hours later he returned in an excited state and said he had seen men in jungle fatigues and wearing caps with red stars on them. All hell broke loose, and we all assumed we might come under some sort of attack. Of a station defence plan there seemed none, but a company of the Malay Recce Regiment soon appeared, led by Lieutenant Aji Singh, who spoke English better than most of us. A sweep through the nearby jungle revealed nothing, but the MO thought Johnson to be rather strange and on probing it was revealed that, far from being a steward, he was an airframe technician. On being asked why he pretended to be a steward, he said that he had always wanted to be a barman. Not surprisingly, Johnson disappeared to Changi Hospital and was never seen again.

The supply services offered by my team were very basic, and besides clothing and some barrack furniture, the main activities were laundry for both the station staff and the detached unit and aviation fuel for the resident six Canberras. As a flight commander there was little for me to do, although I tried to make the fuel supply for road vehicles slicker and was able to rent a fuel tank to save the almost daily trek into the nearby town by a veritable convoy of service transport.

I did have one staff issue and was annoyed by it. I learned that a replacement from the UK was being retained at RAF Tengah and an SAC from that unit would be sent to me instead. I was quite naive still and the thought that someone would unload their problems on me, had not crossed my mind – stupid boy! The man eventually appeared, and I spoke with him about being sent to a detachment in so basic a place as Kuantan and he told me that he knew he was being 'got rid of', not for any disciplinary issues but because he was involved with a bar girl, and although he knew better, it was supposed she was a lady of easy virtue.

As mentioned, the flying squadrons deployed to Kuantan very much looked after themselves and with the supposed three-month detachment for the rest of us, people came and went all the time. The first detachment CO was a navigator, and he was replaced by Squadron Leader Bernie Spry who came from a Hastings squadron at Colerne, when that element of the transport force was stood down following a major accident near Abingdon. In the meantime, I was working my way up the 'seniority' ladder for repatriation to the UK. With air transport at a premium, and the RNZAF Freighters working hard to fill the gap, the supplies being ordered were coming by road up the east coast from Singapore, with the airspace reserved for freight, of the self-loading variety. The adjutant was a fighter pilot, and he topped the pack for ingenuity when it came to leaving. He seemed very confident about leaving, even though no passenger flight was planned for that day, and he certainly wasn't going to pay for a civil airline ticket. With his bag packed he bid us his farewells and strolled towards the loading point. At that stage a Meteor T.7 appeared in the circuit and landed – the clever clogs had pulled some strings and a two-seater from the Changi Station Flight was sent to pick him up. We were treated, however, to an above-average beat-up when the Meteor departed.

Near the end of my stay, I was given a break to take a piece of broken communications kit to Singapore for replacement. With the RAF 'airbridge' still fragile, I was told to go by road and was given a Land Rover and an armed escort. As we were travelling in part over the weekend, I was authorised to stay an extra night before return. We set off in the late afternoon and made good progress before dark, crossing several rivers on the way until we reached Endau, which required a ferry crossing. The ferry had a capacity of four or five cars and after a short delay we boarded. There then came a potential problem when it was realised a car in front contained four Japanese businessmen and tempers flared from the Malays who, just twenty years after the war, still had memories of the occupation.

We continued south along the east coast road and beyond Mersing inland through central Johore and over the causeway to Singapore. The broken crypto machine was handed in, our firearms lodged in the armoury, and I then got my little group some accommodation and told them about the return journey. After more than three months of indifferent showers, the thought of relaxing in a hot bath was most welcome and I took full opportunity of it.

Although I found my first overseas stint to be interesting, by the time I reached the three-month point, I was ready for pastures new and I was a little 'put out' to find that the officer due to replace me had been delayed and hence several more weeks were ordained. Eventually, an officer with whom I had trained turned up and with much the same handover as I had had, I took my departure but not before experiencing the joys of ninety minutes in the full volume of the RNZAF Bristol Freighter sans soundproofing of any sort. I have to say I spent the whole time looking at the clamshell cargo doors, expecting them to open at any moment and deposit us into the jungle below.

For the third time I found myself at the Changi Creek Transit Hotel with a few days to spare before my return to the UK. On the second morning, I was 'paged' by reception and told to report to HQP&SS at HQFEAF and it was a somewhat apprehensive pilot officer who turned up wondering whether I was to be sent home in irons or whatever – the fact that I was met by a provost officer who smiled, only heighted my concern.

I was told that there had been a string of unexplained deaths amongst the dogs at RAF Changi and organ samples were to be sent to the UK and I was to act as custodian of these samples and at each stop, the dry ice would be replenished until at Lyneham the sealed canister would be taken from me.

I was to be picked up from Changi Creek and driven straight to the aircraft immediately prior to its departure. Whilst this treatment had a certain VIP feel about it, being escorted up the aircraft stairs by a burly police NCO and at the en route stops having an equally burly police NCO enter the cabin and call my name out and escort me from the aircraft first, must have suggested that I was in some sort of deep trouble and I noted that nobody seemed to want to sit near me when refreshments were offered in the lounges.

Not long after my return a signal was apparently received posting me to the RNZAF survey team based in the Falkland Islands and this before most people knew where the Falklands were. Some weeks later, I received a formal posting notice, which was much more to my liking – how could I not like three years at RAF Nicosia!?

I contacted the officer I was to replace, and we exchanged several letters but then through the service mail, I received two Forms 4961, (for the uninitiated: 'Posting and Travel Instruction Overseas – Officers'). The first cancelled my posting to Cyprus but the even worse news was that I was to go back for a one-year unaccompanied tour to RAF Labuan, Borneo. As I was dined out from Tern Hill, the supposed job in the Falklands was revealed as a spoof by a wing commander, who I had hitherto thought had no sense of humour. So, less than a year since I had climbed the steps to a British Eagle Britannia at Heathrow, I was doing the same again (as I would fly several more times in G-AOVB it seemed like déjà vu).

On arrival in Singapore, it was once more to the Changi Creek Hotel and a chance meeting with three helicopter pilots from 230 Squadron who were on their way back to the UK, having finished their one-year tour. The last time we had met was in the same place as we parted as they started their sojourn.

I once more presented myself to HQFEAF, but this time more appropriately dressed and with a slightly smug feeling that I knew my way around. The following day I was off for a four-hour flight by Argosy (the Whistling Wheelbarrow) and a decent handover from Flying Officer Vernon Munday before he headed back to the UK.

As soon as possible I tried to get some flying in and within a few weeks was on an eight-hour anti-piracy sortie around the northern part of Borneo and I also found an excuse to go by Twin Pioneer for a day into the interior of Sabah, as well as Brunei.

I found myself inheriting the job of films officer from Vernon and this required me to go to the docks in Victoria, the main town on Labuan, and collect a stack of 16-mm films from the chief officer of the weekly ship which brought most of our stores. These films were then circulated during the week and returned to me to be replaced by the next lot of films. The SSVC (Services Sound and Vision Corporation) ran courses for those who were projectionists and I immediately saw an opportunity for a week in Singapore doing the necessary training course. Sadly, it was ruled that all I needed was a quick refresher and hence no week away.

Two Whirlwinds of 110 Squadron at Nanga Gaat in Sarawak.

About three months after I got to Labuan, the 'war' was over and there was a rapid rundown and repatriation of personnel. When the drawdown had been started, the equipment staffs at HQFEAF insisted there was to be no equipment written off and everything would need to be accounted for and recovered, even from the remote areas. In my time at both Kuantan and Labuan, I never saw any staff officers from Singapore, other than a fleeting visit as the senior equipment officer passed through. After four years of Confrontation, and with staff turning over at least annually, expecting there to be no issues was unrealistic. We carried out a thorough stocktaking exercise but there were some significant errors, many of which could be attributed to record keeping. I was then sent to Singapore to get the costs of the discrepancies, so we could prioritise our investigations. Somehow, the HQFEAF staff learned I was at the depot in Seletar, and I was told to report to HQ 'chop-chop'.

I explained what I was doing but it seemed to me that the staffs were trying to suggest that any problems were caused by the current incumbents, and I was ordered not to bother with prices of kit but to return to Labuan and find the stuff. The parting shot from Wing Commander Bill Groves was that he would follow my career, what was left of it, with interest. I returned to the depot with considerable despondency but was told by Squadron Leader Bill Fogden what he could do to help and how I could resolve any issues. With a burden lifted off my shoulders, I returned to Labuan, and this was the first of many times in my service that I had cause to be grateful to Fogden.

One outcome of the drawdown was a decision to move a detachment of Whirlwind helicopters from Kuching to Labuan, to undertake the residual tasks which the RAF would

Landing Pad Yellow 177 below the border ridge with Indonesian Borneo.

Refuelling from drums on the log landing pads at Long Somado, Sarawak.

carry out. A signal was received at Labuan and was given to one of the SNCOs, responsible for ordering and arranging 'packed fuel' stocks (i.e. 45-gallon drums), so that we could preposition stocks in our area for the move. When I asked the NCO to confirm the agreed arrangements, he confessed to have forgotten. I went into panic mode and although the Shell manager could supply the fuel, he could not deliver it to the airfield at Lutong/Miri, which was a coastal strip just inside Sarawak. I was able to enlist the help of the RAAF training officer on the RMAF Caribou squadron, but when the RAF movements officer learned what I was doing he gave me some sound advice on carrying the drums in an aircraft and had his people on site to help load and secure the drums. I have always been grateful to Ron Leonard for his counsel.

With 110 Squadron firmly established, and the pressure somewhat more relaxed, I was able to do lots of flying, some for business but a lot for pleasure. On one occasion I spent a whole day flying around the airstrips and jungle forts, searching for a paraffin-powered fridge which was the property of 110. It was eventually located in a locked compound at Brunei airport, and having persuaded the Gurkha sentries to help us, it was restored to its rightful place in the crew room, fuelled by some spare avtur which just happened to be lying around.

Christmas 1966 inevitably found me as duty officer, but the seasonal feeling was maintained and I was on the downhill leg to repatriation. A most agreeable conclusion was the offer of two years at RAF Kai Tak – but that's another story.

CHAPTER TWENTY-ONE

WHIRLWINDS IN BORNEO

PETE CHADWICK

Unlike the current crop of young pilots joining the RAF, who can take many years to complete training, I was on my first operational squadron well within two years. This was due almost entirely to the build-up of the Indonesian Confrontation in the early 1960s, which could be considered the United Kingdom's equivalent of the United States' Vietnam. Prior to this point, RAF helicopter squadrons were manned largely by experienced ex-fixed-wing pilots. However, the increasing intensity of rotary operations in Borneo triggered a steep rise in helicopter pilot requirement and resulted in the introduction of ab initios to the role. This was achieved, initially, by pushing basic flying training graduates through an advanced flying training course on Varsities and thence to a rotary flying course. In the event, even this did not satisfy the demand and it was decided to further shorten the progression and I was among the first cadre of student pilots to go directly from getting 'wings' on the Jet Provost to helicopter training at RAF Tern Hill without any advanced flying training.

Thus, I found myself posted to 230 Squadron one year and nine months from joining the RAF at South Cerney. The squadron was then in Germany, equipped with the Westland Whirlwind HAR 10, but was due to move en masse to North Borneo, in early 1965. The squadron returned to RAF Odiham, regrouped and then

Peter Chadwick with the 230 Squadron sign at Labuan.

we flew all our aircraft on to HMS *Triumph*, an old wartime carrier converted to an engineering/factory ship and, with a couple of pilots and technicians on board, which sailed for Singapore.

The rest of us departed as a squadron on board a British Caledonian Airways DC-6, also bound for Singapore, in late February 1965. The journey took two days and involved a total of six legs. As the junior officer on board, I was still an acting pilot officer. I was appointed officer i/c troops and required to submit a report at journey's end on the performance of the civilian crew in looking after the passengers. Needless to say, I was well looked after during the trip.

Whilst in Singapore, we were based at RAF Seletar. When *Triumph* arrived, we flew the aircraft off and spent some time sorting them out. Air tests, compass swings, blade tracking etc., before some basic acclimatisation flying into Malayan jungle clearings. The Whirlwind HAR 10, a Westland-built version of the Sikorsky S-55 or H-19, a single-turbine engine-powered single-rotor helicopter was, we discovered, quite limited in performance in the hot, humid, and high conditions of the Far East when compared to what we were used to back home. It was not unusual to be trading fuel for payload, or vice versa. To be able to carry the equivalent of five or six soldiers a respectable distance was sometimes a problem.

Those of us who were going to be flying over the Borneo jungle, which was most of the squadron, undertook a truncated two-day jungle survival course, to prepare us should we ever be forced down in Borneo. Apart from the pilots, our aircrew consisted of squadron technicians who had been trained on the squadron and were paid a pittance of a daily rate of 'flying pay' to act as rear crew. (This was before the days of air quarter masters, subsequently air load masters, who started to filter through to the squadron from the fixed-wing transport fleet sometime later that year.)

At Seletar, we were fascinated by a wonderful carved, wooden 230 Squadron badge in the entrance hall of the officers' mess. This dated from the time between the wars when 230 had been operating flying boats from Singapore. The badge is of a tiger standing before a palm tree, which, rumour has it, was taken from the logo of the Tiger Beer brewery of Singapore when squadrons were required to select their own squadron badge during the 1930s. We lost no time in advising the Tiger Beer brewery that 230 Squadron was back in town, which elicited an invitation to visit the factory and have a guided tour. An invitation which was eagerly accepted.

Once this short acclimatisation period was over, we embarked the squadron, lock stock and barrel, in HMS *Bulwark* and sailed for Labuan Island off the north coast of Borneo, where we flew off to our new base. Labuan was not large, but it was quite a busy airfield used by Borneo Airways Dakotas; the RAF, with Argosy and Beverley transport aircraft, Single and Twin Pioneer short take-off-and-landing aircraft; the Royal Australian Air Force with Sabre single-seat fighters, and the Royal Malaysian Air Force operating Handley Page Heralds. In addition, we had regular visits by Singapore-based Hunters, Javelins, Hastings, and Shackletons. Luckily, we occupied our own little enclave on the other side of the single runway from the main dispersal; meaning that we could usually depart the airfield unhindered by any ATC delays.

Our operating area was to be the whole of Sabah (North Borneo), with forward operating bases (FOBs) at Tawau, a small airfield in the extreme south-east of Sabah; Sepulot, a small grass strip a few miles north of the Indonesian border in central Sabah; and Bario, a similar grass strip in the extreme west of Sabah. The main task was to be supporting, deploying forward, and resupplying Gurkhas, Guards Para, and New Zealand and Australian SAS at their remote operating bases.

The first task, though, was area familiarisation. There were no topographic maps. All our maps were simply of river courses with no information about what lay between. We spent some time becoming very familiar with certain distinctive trees at the confluence of several rivers, high ridges and unusual hill or rock formations.

Fuel dumps were at several unmanned remote sites. In the first event, fuel was simply in barrels with hand-operated 'wobble' pumps for transferring avtur into the aircraft. Latterly, we had the luxury of Kelston motorised pumps with two-stroke engines which was much less labour intensive. Initially, when we could afford the weight, we flew with squadron corporal technicians as crewmen. This meant that, firstly, we had a decent chance of fixing any aircraft faults and secondly and probably more importantly, they could help with the wobble pump or starting the Kelston.

Once we were deemed to be area familiarised and operational, which took less than three weeks, the detachments of 110 and 103 Squadrons were relieved, returned to their main bases at Seletar, and we assumed responsibility for all North Borneo.

The general pattern of flying was that we would spend two or three weeks at one of the FOBs, followed by two weeks or so back at Labuan. At the

Approaching a manned fuel dump at a jungle dispersal.

FOBs it was a seven-day operation throughout, sometimes spending one or two days on forward tasking, sleeping in or by the aircraft, or, if the native Muruts or Dayaks were particularly friendly, in the local village long house. When we were back at Labuan we would check the tasking first thing and if we weren't required for operations or some flying training, we would have the day off. Perhaps to go down into Victoria town for a haircut; a local entrepreneur had assessed the market and established a barber's shop staffed by topless lady barbers; our hair was always neat, tidy, and short when back at Labuan. Or we simply took a break away from constant duty.

Often, we would fly without crewmen to save on workable payload. (Gurkhas were masters at self-embarkation and sneaking on an extra man whilst the pilot turned away to take off.) We had been issued with flak jackets (bulletproof waistcoats) but they were heavy and cumbersome – and common sense dictated that any bullets coming our way would most likely come from below – so, they were rarely worn but folded up and sat on. We each had a Stirling sub-machine gun, complete with two full magazines, which were generally carried laid on the front coaming, and a 0.38 revolver. The latter was a little pointless for me as I couldn't hit a barndoor with it even if I was leaning against it. We also had an 'escape tape' as the perceived wisdom was that, if we had an engine failure whilst over primary jungle – remember we had only one engine – we would probably end up in the upper branches of a very tall tree, some several hundred feet above the jungle floor. The tape was to affect an abseil to the ground, but I was never truly convinced.

Accommodation out at the FOBs was rudimentary; at Tawau which was a small airfield with a tarmac runway, which could take the Herald, and where we worked alongside a small detachment of RMAF Alouette helicopters, we had a single-storey mess building, which served as dining room and anteroom, with tents for sleeping quarters; at Sepulot and Bario, which were grass strips able to take Twin and Single Pioneer aircraft, we had a corrugated iron basha or longhouse for sleeping and an open-air dining room/lounge area.

Back at Labuan, in relatively civilised circumstances, we initially had the use of the airport hotel, which had been requisitioned as an officers' mess at the start of the Borneo operation, and the Membedai Club, which had been the Shell Oil social club in its day, also requisitioned for the purpose. The requisitioned airport hotel was straight out of the 'British Raj' with no air conditioning, mosquito nets in bedrooms, big ceiling fans and chitchat lizards living on the ceiling, whereas the Membedai Club was much more comfortable with air-conditioned accommodation and all modern conveniences. When we first arrived at Labuan, all the aircrew operating from there were on detachment and therefore were relegated to the hotel, whereas the non-aircrew support staff, with longer tenure, were comfortably settled in the Membedai Club. Our squadron boss, Dougie Thomas, argued that permanently based aircrew should not suffer the relative privations of the hotel, but should be accommodated in the more comfortable Membedai. This, of course, fell on deaf ears with the hierarchy.

However, shortly after we arrived at Labuan, the then president of the mess committee of the officers' mess was at the end of his tour and returned to the UK. There was therefore a need for a new one. I'll leave it to your imagination as to who volunteered his services and of the subsequent accommodation arrangements for 230 Squadron aircrew.

Our part of the island of Borneo was North Borneo, which had become Sabah, a province of Malaysia. The other part of the northern section of Borneo belonging to Malaysia was Sarawak, and on the north coast, between the two provinces was Brunei, a Crown Protectorate. The rest of Borneo was the state of Kalimantan, a part of Indonesia. The border between Malaysia and Indonesia was, for the most part, a mountain ridge rising in places to 7,000 feet. The confrontation took place across this ridge with both sides making incursions into the other's territory. Along this ridge, UK Royal Engineers had created a series of helicopter landing zones by blowing a gap in the jungle canopy and constructing helicopter landing pads from the felled trees. Our regular job was delivering troops, usually a platoon, to these remote sites on the ridge. The troops would then 'patrol' on the border, and we would either resupply them or recover them back to their bases on completion of their patrol one or two weeks later.

Apart from the routine tasking of delivering or resupplying troops, there were other tasks which came our way. The local natives, Dayaks nearer the coast, or Muruts living in the centre, had no knowledge of the political machinations going on between Indonesia and Malaysia. And to be brutally frank, I doubt that they would have concerned themselves about which they lived in. They lived in very remote villages, usually beside or close to a river, consisting of one or two longhouses, a central grassy area, known as a Padang, and very often a pristine and carefully manicured football pitch. Occasionally, we would be tasked to take a small party of civil servants to several villages on a 'hearts and minds' visit to promote the value of being Malaysian. Whilst the civil servants were going about their business, we were free to look around the village and 'socialise'. I can recall looking into the local schoolhouse to witness children, who lived in a village, miles from any semblance of a road, being taught the Highway Code.

One of the traditions of the Murut was that friendly visitors were celebrated and invited to refresh themselves with some of the local brew, Tapai; in fact, to refuse the refreshment was considered something of a slight. Tapai was a very strong alcoholic brew of tapioca leaves mulched into a large pot, topped up with questionable river water and allowed to ferment for some time before being siphoned off for consumption. We didn't generally visit unannounced, and because we were known in the villages that we did visit regularly, as being 'on duty', we were allowed to skip this ritual.

Whirlwinds at the Sepulot dispersal.

However, there were occasions when visits to villages not familiar with this concession were forced upon us. I can remember one occasion when transiting to Sepulot from Labuan, fortuitously with another, more senior pilot, being caught by a deterioration in the weather forcing us to land at a remote village en route whilst the weather passed. It was clear that the locals accepted us as friendly, although unusual and unexpected, visitors and invited us to take 'refreshment'. I'm not sure how we did it, but we managed to persuade them that only one of us could enjoy the refreshment, because the other had to fly the helicopter once the weather cleared. Since I was the junior pilot on board, you can guess who had to accept the hospitality; suffice to say that my memory of the subsequent leg to Sepulot is a little hazy.

On another occasion, whilst on detachment at Bario, I found that I and my crewman, Flight Sergeant Eddie Beacon, had been tasked to take a load from Bario to Seria in Brunei on the north coast. On arrival at the aircraft, we found that the 'load' was a water buffalo calf, alive and kicking. Some investigation revealed that the annual Gurkha festival of Dussehra, which involved the ritual slaughter of a water buffalo, was fast approaching and it had been determined that obtaining a suitable candidate was significantly cheaper up country than on the coast.

We spent a little time discussing how we were going to achieve this task and how we were going to carry the load. One possibility would be to carry the beast below the aircraft hanging in an underslung net. This was a well-practised procedure to carry inanimate loads such as ammunition boxes, or fuel drums, or even hard core for road building, but we didn't feel that this should be visited on the poor unknowing water buffalo, as this would mean it hanging free below the aircraft for a trip of something only a little short of 100 miles i.e. for an hour or so. The upshot was that we coaxed the beast into the main cabin up a ramp, tripped it onto its side, and lashed it down with nylon load tape. Eddie accompanied it in the cabin during the trip, with his .38 revolver at the ready.

On arrival at Anduki Airfield in Brunei, which was on the outskirts of Seria, we were met by the adjutant of 2/6th Gurkhas, whose load this was and who asked us to deliver the animal to their HQ in central Seria. Since we were unfamiliar with this destination, he offered the possibility of leading the way in his British racing green MG. Suffice to say that the rather bizarre circumstance of a Whirlwind helicopter carrying a live water buffalo calf, following an MG sports car to the 2/6th Gurkha HQ in Seria, was completed successfully; and we found ourselves in possession of a significant amount of Tiger beer to take back to Bario.

On another occasion, when back at Labuan, I found that I had been tasked to pick up the local British General Officer Commanding and his adjutant, from his residence in Brunei and take him on a tour of some of his forward troop bases in the lead-up to Christmas. Bear in mind that I was a very newly promoted substantive pilot officer and my recollection of the sortie was that I knew that I was carrying a very senior army officer, but I can't for the life of me remember his exact rank. I would guess, looking back, that he was two or three star. Since I knew our patch pretty much like the back of my hand, I just took him and his aide-de-camp on a guided tour of his domain, dropping in on his forward positions, as required, under the ADC's direction. This entailed lots of landings in clearings and FOBs and quite a bit of sitting idly by whilst the visits took place. Towards the end of the day's tour, we landed at a landing zone a short distance over the ridge from Sepulot. So, since we now had insufficient fuel to return to Brunei direct, I took the opportunity to transit there for fuel, whilst the general was visiting his troops before returning to pick them up and take them back to his residence in Brunei. Having shut down in Brunei and disembarked them, the ADC returned to the pad and indicated that he wanted to have a word. He climbed up the cockpit to speak to me and said, "The general would like to know why you don't do as we do in the army and refuel your vehicles overnight?" It must be remembered that this was still in the early days, for some, of the army operating with RAF helicopters.

The tour on 230 Squadron in Borneo was twelve months unaccompanied. Not a major trial for us young single chaps, but not so easy for the more senior, mature and generally married members of the squadron, whose wives were left to run and manage the family home for the whole of that period. During the twelve months there was an opportunity for two weeks' leave at about the six-month point. It's probably true that all the young single chaps, without exception, spent this leave in Bangkok. Bangkok in the mid-1960s was a wonderful place to take a break from a year during which there had not been too much female company. But any more detail about this break in the tour is probably best left unspoken.

During my year in Borneo, I flew a multiplicity of troops in very varied situations; New Zealand SAS, Australian SAS, Guards Para, and Gurkhas. Of them all, it was the Gurkhas who I mostly admired; willing, enthusiastic, fiercely loyal to the Crown and to their officers, and for all their professionalism, the most fun loving.

I look back on the tour as both an enjoyable baptism of fire and a very steep learning curve, with a need to simply get on with the job largely by one's own initiative, relatively unsupervised on a day-to-day basis. We had no other duties, no orderly officers, no station duty officer, in fact no ancillary duties at all except to fly and get the job done. It was marvellous.

CHAPTER TWENTY-TWO

LIFE ON A HASTINGS SQUADRON BASED AT CHANGI

TONY STEPHENS

THIS IS A BRIEF PERSONAL account of my tour as a first tourist navigator on 48 Squadron, Changi, between 1963 and 1966. After six decades the memory fades and my recall has been refreshed by consultations with colleagues and with reference to RAF histories, as well as to 48 Squadron's ORBs/Forms 540, my flying logbook and personal notes.

ARRIVAL

First impressions of Singapore were not favourable. Exhausted, dishevelled, travel stained and 'jet lagged' after a twenty-four-hour flight with British United Airlines, on descending the steps at Paya Lebar, Singapore Airport, we were met by a wall of steaming tropical heat. This discomfort was only relieved after a few cold Tiger beers in RAF Changi's Temple Hill officers' mess (one of the more pleasant spots on the island), which was to become my home for the next couple of years. Very few lucky enough to be posted to the ageing Hastings of 48 Squadron in Changi, Singapore, realised that it was probably going to be one of the most rewarding and enjoyable tours of their RAF careers.

CONFRONTATION

We walked straight into a controversial political situation which was to last throughout my tour. The politics of the area covered by a divided Borneo reached a climax with the

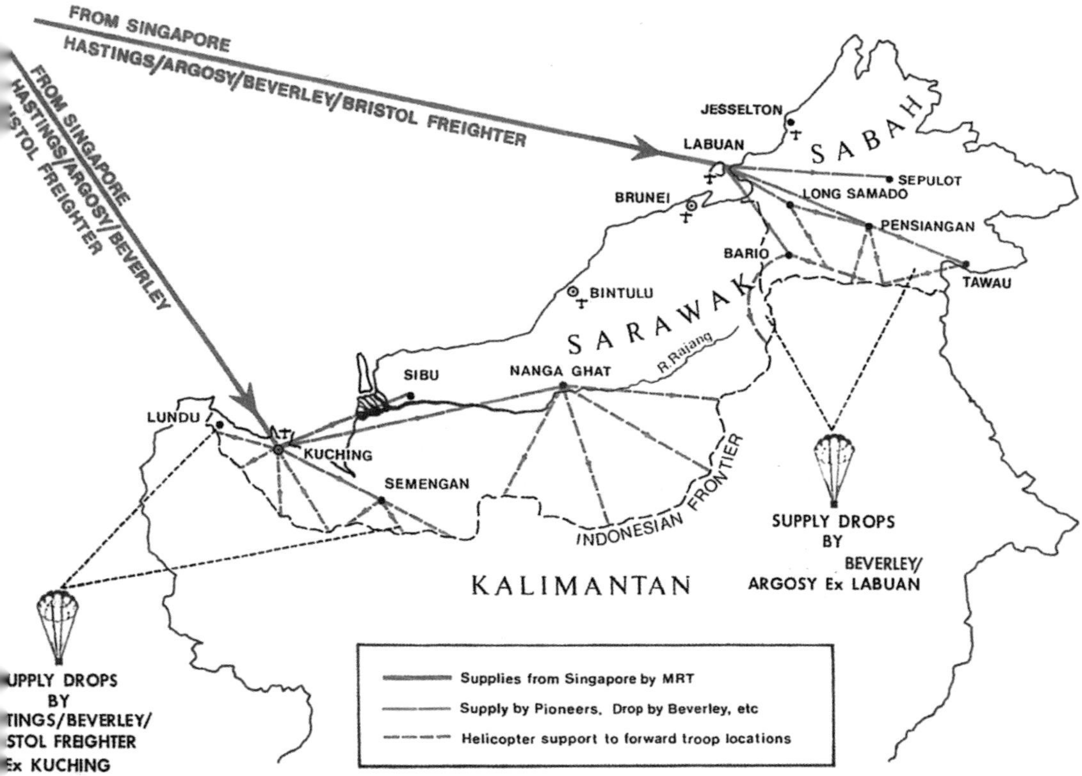

Fixed-wing transport operations throughout Borneo.

formation of the Federation of Malaysia on 15 September 1963. The roots of the problem lay in President Sukarno of Indonesia's objections to the new state incorporating Sarawak, North Borneo (now Sabah) forming an eastern division with Malaya and Singapore the western division. Diplomatic relations with Indonesia had been broken off, and a state of undeclared war, to be known as 'Confrontation', existed. Indonesian cross-border incursions into Malaysian Borneo increased and it became necessary to reinforce the few army units in some of the most inhospitable and inaccessible jungle areas on the border. Although helicopters provided invaluable support to these forward units, major resupply was to be by fixed-wing aircraft; the 48 Squadron detachment was based in Kuching.

48 SQUADRON

Equipped with some twelve aircraft and eighteen crews, 48 Squadron's main role was to provide air support to the Far East Land Forces. This involved not only supply dropping duties in Borneo, but also carrying passengers and freight throughout the theatre. The most frequent schedules were to Labuan, Kuching, Hong Kong and Gan. It also contained a detachment of two aircraft and crews to Christmas Island, from where they flew in personnel, fresh food, equipment, and mail from Honolulu.

A Hastings of 48 Squadron over Singapore.

The Hastings had the reputation of being a 'maid of all work' and a robust 'workhorse'; its roles also included casualty evacuation, VIP and occasionally search and rescue flights. With a crew of two pilots, navigator, signaller, engineer, and air quartermaster (AQM), the aircraft was powered by four Bristol Hercules engines and had a maximum payload of some fifty passengers or nine and a half tons. Not being pressurised, it went almost everywhere in its route-flying role at an air speed of 200 kts at an altitude of 10,000 ft. However, with its tailwheel, it could bite pilots not accustomed to its vagaries, especially when landing in a crosswind. It had an astrodome for the bubble sextant and a bay with a Perspex window underneath the cockpit from which the navigator carried the final stages of supply dropping. In FEAF the Hastings had only basic navigation aids: the bubble sextant, a Marconi automatic direction finder (ADF) radio compass, which tended to home in on the nearest cumulonimbus cloud, a Rebecca/Eureka radar for final approaches to RAF airfields and a drift recorder. Thus, for many of the long flights, over both Indian and Pacific Oceans, the navigator had to rely mainly on astro navigation and the drift recorder.

My low-level acceptance check was a typical sticky two-hour cross-country over the Malayan jungle. In such clammy conditions I was grateful for my lightweight flying suit and headset. My route initiation was a scheduled flight to Kimpo, Seoul, Korea to carry out the roulement of British army troops working as part of the UN Force. The route took us through Saigon, at a time when US forces were just starting to build up in Vietnam, and

Hong Kong. Because the approach to Hong Kong's Kai Tak airport was unusual, it was squadron policy for all new crews to be screened on their first flight. On the final approach it was a bit disconcerting to head towards the checkerboard painted on the sheer cliff, only turning sharply at the last minute to let down low over blocks of flats to land on the runway. I also flew my first of several flights to Gan, some eight hours – over little but the blue Indian Ocean. This was the first of many flights on which I was almost totally reliant on the sextant. This always kept me very busy, but much at home with my thirty-minute sun-shot fixing cycle.

CHANGI

Back at Changi, for comfort, on arrival we had all quickly exchanged our heavy-duty UK KD uniform for local KD, made almost overnight by Chinese tailors in the local Changi village. Our working day started early, and if not flying, afternoons were for leisure, or siesta. However, flying commitments, often with early departures, dictated that the nature of the squadron working day was varied. I do recall that the air-conditioned squadron crew room was rarely very busy.

Temple Hill mess was the home for single, unaccompanied, and visiting officers with a mix of RAF, WRAF officers, civilian teachers, nurses, Malcolm Club managers and some army and Royal Navy officers frequenting it. Several met their future spouses in Temple Hill. The common thread among mess members was youth and lack of seniority. This resulted in a vibrant community and a great *esprit de corps*. If not flying, I would swim or water-ski at the local officers' club. Also, we often took our old 'bangers' of cars down the bustling centre of Singapore in Orchard Road, to experience the completely Chinese way of life in Chinatown, or eat out at the open stalls of Bedok Corner. The opportunities for leisure activities were endless.

My home was one of the chalet 'bashas' which surrounded the hill on which the main mess stood. Fans kept us cool and helped to keep the mosquitoes at bay. Deep monsoon drains surrounded the mess. As well as getting rid of the heavy rains, they also housed rodents and snakes. Pythons and cobras were still to be found on the station.

RAF Changi had over 1,500 personnel of all ranks, many with families. There were several clubs covering a range of leisure activities and RAF Changi had its own broadcasting service (CBS). The warm climate was ideal for outdoor sport, with many team games played under lights in the evening to avoid the heat of the day. I was immediately very involved with station rugby. As well as many local games, I enjoyed two tours to Hong Kong with the FEAF team, one as captain, and a tour to Western Australia with the Joint Services Far East Team.

KUCHING

My two detachments to Kuching for resupply-dropping duties were hectic. On the detachment's early operations, which started in July 1963, the crews found it necessary to fly the Hastings to its limits in dropping supplies to fourteen dropping zones, often amounting to more than thirty drops a day in difficult terrain and in exacting weather conditions. Low cloud, poor visibility and heavy rain added considerably to the interest of such supply-dropping flights. Everyone who flies in a Hastings as it goes around mountain sides and up narrow valleys is amazed that such a heavy unresponsive aircraft is capable of such energetic manoeuvres. On another detachment, a Hastings was fired on by Indonesian soldiers with small arms. These extracts set the scene for my detachments, the first being early into my tour. As a result of the earlier experiences, we were now issued with bulletproof flak jackets and a .38 revolver which we always carried on supply-dropping sorties.

The first day was to set the pattern for our supply-dropping operations over the next fourteen days and on future detachments. With the equatorial climate and the frequent proximity of the Inter-Tropical Convergence Zone the weather was always a major factor. Early morning the cloud base was often 'on the deck' so it was not possible to take off much before 09:30 with cloud only lifting slowly as the warmth of the sun dispersed it. Even then, we often had to stay below cloud. Because they had to work through a side door and the aircraft floor had a steep slope, the willing despatchers from the Royal Corps of Transport found loading difficult. Loads were mainly harness packs to be dropped from

Air drop over Borneo.

800 feet; very occasionally we free-dropped sacks from 50 feet. As well as fresh rations and other consumables, loads were an eclectic mix of mail, ammunition, bundles of barbed wire, metal picket posts and occasionally live chickens. The packs filled the aircraft to the roof, such that the crew often had to climb over them to reach the cockpit. With the intense humidity, it was always very sticky.

At this time there were no official topographical maps of Borneo available, and maps were passed from crew to crew; the navigator's map was a blank green with some hand-drawn hachured distinctive features, such as mountains, rocks, villages or clearings. Much of the flying was over featureless jungle. We flew two resupply sorties a day: one in the morning and another in the afternoon, when the warmth of the sun had lifted the cloud base; the convection over the jungle meant that the latter were always more turbulent. On our morning flight we dropped a record 14,000 lbs of supplies to four DZs on the Indonesian border close to Kuching on a flight lasting three hours; it included an SAS DZ, situated on the side of a cliff which demanded skilful flying. Some DZs were postage-stamp sized, while others were more accessible. DZs were usually marked by a dayglo orange balloon which floated above the thick, high jungle canopy. Without this approach aid it would have been almost impossible to locate many of them. Once close, the DZs were identified by a letter – the aiming point – and once ready the DZ party would fire a green Very light. Radio contact was attempted, but not always established. The co-pilot was the lead map reader, but with little to go on navigation was mainly by dead reckoning and stopwatch. Luckily, but not invariably, the winds were often light. As all jungle looked the same, it was very easy to fly up the wrong valley; with the hills either side closing in, having to sharply turn 180 degrees in a Hastings was challenging to say the least. Teamwork was paramount and we needed all eyes in the cockpit to locate some DZs. As we approached the DZ, I eased my way into a prone position in the bay for the drop. Lying on bulletproof flak jackets, I had to rely on chinagraph vertical and horizontal lines on the Perspex window for initial guidance on when to drop, calling the red ready light and green release lights to the AQM and despatchers. We did try to avoid dropping outside the DZ and into the jungle, but, for various reasons, were not always successful.

The afternoon sortie on this first day was a much longer flight lasting four hours which took us very low level along the Rajang River before climbing on reaching the mountains to fly above cloud and then descending through a gap to the Gurkha DZ at Long Jawi, well upcountry. This sortie was memorable, as a bundle of pickets broke open in the drop and the AQM reported that he saw all these individual 'spears' of metal hurtling down to the DZ on what was one of my more accurate drops, scattering the Gurkha DZ party.

Towards the end of the detachment an ADIZ was established over Borneo and we enjoyed fighter escort in the form of a Javelin from Tengah on some sorties. It had been an

A precision drop of supplies.

exhausting and demanding, but very satisfying couple of weeks, in which we had dropped some 100,000 lbs of supplies. A year later, on my second detachment we flew to some thirty DZs. Most sorties were just one to two hours long, with the occasional three to four hours upcountry. On this detachment we dropped some 190,000 lbs. On one drop a crate of chickens broke up as it left the aircraft, leaving the poor birds, which had never previously flown, plummeting down to the jungle below. Navigation this time was helped by the provision of a proper map, but the stopwatch was still invaluable.

CHRISTMAS ISLAND (NOW KIRITIMATI ISLAND)

After the hectic activity of Kuching, my two Christmas Island detachments which followed were relatively relaxing. Getting there from Singapore now involved a four-day flight as Confrontation had necessitated flying around, instead of across, Indonesia, and staging through the remote Australian-administered coral island of Cocos in the Indian Ocean.

The route then took us to the RAAF base at Pearce, Perth in Western Australia, and on this occasion across the vast yellow and red deserts of central Australia on a nine-hour flight to RAAF Amberley, thence to a welcoming green Fiji, before crossing the International Dateline to Christmas Island. Setting forth on long trips from Changi in the Hastings always had the potential of being an adventure, and rarely did we avoid having aircraft problems of some kind. As the CO commented: "the vintage old ladies are possibly beginning to feel their age." However, on this occasion we had no problems.

RAF Christmas Island was established in 1957 for the Operation Grapple nuclear tests and retained for US nuclear tests in 1962. Situated just north of the equator, and some 1,200 nautical miles south of Hawaii, it is a desert island atoll with a circumference of some 100 miles. The detachment's primary role was to provide the station with fresh food supplies from Honolulu.

However, it also had a secondary SAR role for which the station was equipped with an air-sea rescue launch with a local crew. The launch was also fitted out for deep-sea fishing and some detachment members had enjoyed considerable success fishing the virgin waters outside the reef. Other activities to keep us occupied included water skiing on one of the many lagoons which were home to manta rays. Hickham Air Force Base Honolulu was some six flying hours over blue ocean away. At Hickham we parked next to the large USAF transports; of an earlier generation, next to them, the Hastings with its tailwheel looked quaint. The local USAF crews dubbed it 'the four-engine gooney bird' and treated it like a museum piece.

Life at Christmas Island was usually peaceful and fairly laid back, but on Good Friday 27 March 1964 an earthquake of 9.2 on the Richter scale in Anchorage, Alaska, gave rise to excitement and some alarm as it generated a twelve-foot tsunami heading south across the Pacific. Those living near the beaches on the Hawaiian islands, including one of our crews, were evacuated to higher ground as the wave came in over the foreshore. This was of some concern to me as duty officer at Christmas but, in the event, the wave had dissipated on its journey south and was not felt on the island.

Excitement over, with stops in Fiji, Sydney, Perth, and Cocos we enjoyed an uneventful return to Changi. But my stay was not for long as I was soon on my way back to Christmas Island again, this time routing through New Zealand. On arrival, we found the station rapidly running down prior to closure in July 1964. The detachment continued its schedules, but life was basic and there were few leisure facilities. As it was not deemed worth removing equipment and vehicles from the island, much was ditched, although we did rescue two sailing dinghies and some books.

Two events livened up this final detachment. We carried out an emergency free drop of penicillin on to a large Union Jack flying next to palm trees in the grounds of the

British resident on the tiny Washington Island (now Teraina), some 300 miles away. Also, after a fun cricket match, we were invited to a lively farewell evening 'Bartery' in a long-thatched basha by the local Gilbertese islanders, with much singing (including an unusual South Seas version of 'It's a Long Way to Tipperary'), dancing and tub thumping.

Following our now familiar route home, our return to Changi was rudely interrupted on the leg to Cocos when number three engine mutinied after two hours flying over the Indian Ocean. We drifted back to RAAF Pearce. When the engine sump was drained, about six lbs of scrap metal fell out, including one large bolt which had sheared. Following the fitting of a new engine from Changi, an air test and compass swing, six days later we set off on a late-night leg to Cocos which challenged my night astro-navigation skills and star identification. After an exhausting eight hours, I was very relieved to spot the runway lights of Cocos through the murk of a tropical storm.

UNREST IN SINGAPORE

Soon after my return to Changi, our generally calm life was disrupted by the station being locked down for a time because of inter-racial trouble in Singapore. Not long after this local disruption, Indonesian insurgents and paratroopers landed in southern Malaya which prompted an emergency to be declared with RAF Changi taking precautions against a local attack – sandbagging buildings, digging slit trenches, and deploying arms around the station. All leave was cancelled and those on leave recalled; the station went on to a seven-day week for a while, the only relaxation being a curfew between 06:00 to 08:00. However, the squadron was able to continue flying operations. During the month it took to mop up the insurgents, I flew two unusual sorties: a coastal-recce sortie looking for insurgents and a six-hour tedious overnight sortie in which we acted as a UHF radio relay between an RN ship and a radar station.

With the impending rugby season, my flying was now generally limited to local training and Borneo flights. However, route flights included a recce around northern Thailand, to airfields loaded up with USAF F-4s and F-5s for Vietnam, and a twelve-day training flight to sound out a possible schedule to New Zealand. This flight was beset with delays because of the kind of problems facing such an ageing aircraft, mainly fuel and oil leaks. I also flew a training flight to Japan through a typhoon, with rain pouring into the cockpit. In addition, as well as Hong Kong schedules and the increasing number of flights to Labuan and Kuching, which included casualty evacuation and bringing back body bags, I was also involved in a squadron call-out to deploy the RAF Regiment Squadron from Butterworth to quell an abortive uprising in Gan.

GROUNDED

We were all to be brought up short on 6 July 1965 when all Hastings operations were halted following the devastating crash of a Hastings from RAF Colerne when on a parachuting exercise at Weston-on-the-Green near Oxford. Forty-one people were killed including crew and parachutists. This incident led to all Hastings being grounded. In short, the cause was the failure through metal fatigue of two elevator bolts in the tail unit, leading to loss of control of the aircraft. All Hastings had to have their tail units stripped down, inspected, and repaired.

During the two months or so before we could get back in the air again, I taught local employees English and RAF History in the education centre and flew some recce flights over the Malayan coast as an observer in a Single Pioneer.

NORMAL OPERATIONS RESUME

Normal operations resumed some two months later. Once again, my logbook for several months is filled with local training and Borneo flights. One memorable event was that the squadron resumed practising formation flying partly in preparation for a mass fly-past by aircraft from all FEAF squadrons for the Duchess of Gloucester who was visiting Changi.

Formation of 48 Squadron Hastings under turbulent skies.

The FEAF rugby team captained by Tony Stevens (centre).

Unfortunately, the fly-past was planned for 14:00. Right on time, there was a torrential cloudburst which lasted for over an hour. The fly-past was cancelled, but with airfields unusable while we held over the South China Sea, the fast jets were running out of fuel and were pushed to make it back to Tengah and the helicopters were scattered and landed where they could in South Malaya. Otherwise, apart from a training flight to India, and taking the Joint Services Rugby team to Perth, the resupply flights to Labuan and Kuching continued unabated. I flew fifty-five such flights during my tour.

THE PARTY'S OVER

My final months included a farewell flight on a now familiar route to New Zealand and a memorable rugby tour to Hong Kong. As I walked up the steps of the British Eagle Britannia heading home, I was very conscious that the last two-and-a-half years would be unforgettable; it was unlikely that I would enjoy such an experience again. I had flown some 1,700 hours in support of many FEAF operations with some great crews and to many countries from the Indian to the Pacific Oceans. In so doing I had greatly expanded and honed my navigation skills. I had also enjoyed the best rugby of a long rugby career and made many enduring friendships. It was a once in a lifetime experience.

CHAPTER TWENTY-THREE

FIGHTERS OVER HONG KONG

PETER HEADLEY

After graduating from the RAF College Cranwell, I completed the Hunter conversion course at RAF Chivenor and was posted to 28 Squadron in Hong Kong. It was a bit of a surprise to find that the squadron was still flying the de Havilland Venom FB.4s, the last in the RAF to operate the type.

I arrived in Hong Kong soon after New Year 1962 and joined the squadron based at RAF Kai Tak. Squadron Leader Ian Stanway was the CO, flight commander was Flight Lieutenant 'Chunky' I'Anson, and the pilots were David Scott, David Crichton, and myself. The adjutant was Flight Lieutenant Lionel Grindley, an ex-Shackleton navigator. The aircraft were four Venom FB.4s and a Vampire T.11. A mighty fleet to keep the Chinese on the other side of the 'Bamboo Curtain'.

RAF Kai Tak was in the New Territories on the eastern side of Kowloon Bay, past the civilian terminal, as we shared the airfield with Hong Kong's civil aviation terminus. The station commander was Wing Commander J. Knox, soon to be followed by Wing Commander J. Mann. The AOC Hong Kong lived on the island, on the road to Repulse Bay, and the combined military HQ was near HMS *Tamar*, on the north coast of the island.

As well as 28 Squadron HQ, the airfield had aircraft maintenance and storage hangars, all equipped with steel ropes over the roofs and anchored into the ground, to keep them safe during the typhoons which hit Hong Kong each autumn. This was certainly needed later in 1962, when Typhoon 'Wanda' hit the colony with 120 mph winds on the airfield, and gusts of over 240 mph at Lion Rock, in the hills just behind the airfield.

There was a RAF marine section, with a 60-ft air-sea rescue launch. Its main task was to patrol the Port Shelter danger area when it was active for live firing and ensure there were no incursions by Chinese fishing boats. We used the air-to-ground target for live rocket and

Peter Headley (on the nose of the Venom) and fellow pilots of 28 Squadron.

cannon fire. The launch also provided this service to the army guns, who also had a target on an island in Port Shelter. A more social service was to transport 28 Squadron members to visiting US Navy, Australian navy, and Royal Navy aircraft carriers, when invites to onboard receptions were given by these ships to personnel based at RAF Kai Tak.

The Hong Kong Auxiliary Air Force based its helicopter at Kai Tak, which was used on anti-smuggling sorties to help the Hong Kong police and for rescue missions. They were also based at Shek Kong in the New Territories, together with a flight of the Army Air Corps, operating Auster AOP 6 aircraft.

The officers' mess at Kai Tak was a very comfortable two-storey building and each officer's accommodation consisted of a living room and a bedroom. I had a delightful Chinese batman called Wong who looked after his men and their accommodation very well. At the front of the mess was a lawn, where film nights were had for the mess members throughout the summer. The mess was elevated on a hillside and had a good view towards Kowloon and the western part of Hong Kong Island, overlooking the station and Kowloon Bay. The lights of Wan Chai were very beckoning through the ante-room windows. The mess had lots of social activity, with a dining-in night every month, including 'fan cricket' afterwards in the ante room, Curry lunch every Sunday, film nights and entertaining visiting dignitaries such as the governor of Hong Kong, generals, admirals, and members of other services. Other social activities included a RAF Kai Tak rugby team, who played various teams from army regiments, navy ships, banks, and other rugby clubs in the colony. There was a station cricket team, a go-kart club, which had race meetings with local regiments and the station had membership of the United Services Club in Kowloon. The mess also had some interesting animal night life, as one night a mess member was coming back late from Kowloon and had to jump over a very poisonous snake (banded krait), curled up on the steps of the mess entrance.

The daily flying programme started with a single-aircraft weather-check flight around the colony, as, despite its small size, the weather in the north, east and west of the colony could be significantly different to the airfield. It was on some of these flights in 1962, that the tragic evidence of Chinese refugees trying to get to Hong Kong by swimming from mainland China was seen, with many bodies found on the beaches in the north-east of the colony, some having been attacked by sharks.

After a couple of check rides with 'Chunky' I'Anson in the Vampire T.11 on 11 January 1962, I went solo on the Venom the following day and was introduced to an aircraft with different capabilities to the Hunter.

The routine flying programme included battle and close-formation flying, cannon and three-inch rocket ground-to-air live firing, formation aerobatics, army cooperation with the Hong Kong garrison and navy exercises with visiting UK and Australian aircraft carriers. High-level interception training under the control of the radar station at Tai Mo Shan was also carried out. The radar was situated in the centre of the New Territories, on the highest point in the colony.

The high-altitude performance of the Venom was much superior to the Hunter. When visiting RAF and RAAF Canberras came from Singapore and Malaya, they used to act as targets for the Venoms under the control of Tai Mo Shan radar. Flying at 45,000 ft, when the Venom closed in for an air-to-air video attack, the Canberra made an evasive turn. The crew were amazed to find the Venom stayed with them in the 360-degree turns, rather

Peter Headley carries the squadron standard as it is paraded through Hong Kong. (AHB)

Hunter FGA.9s arrive to replace the squadron's Venoms.

than falling out of the sky as the Hunters and Sabres 'down south'. We also acted as low-level targets for the local army garrison's anti-aircraft guns and any Royal Navy ships who were in the area south of Hong Kong. I did this for HMS *Belfast* on her last tour out to the Far East and it was a little unnerving watching the flashes from the gun muzzles, knowing they were pointing at you.

In its old age, the serviceability of the Venom was not good and, on many days, only one aircraft was flying. On these days, we flew mainly for endurance to get as many hours as possible towards the monthly target for the squadron. I can remember one of these when I was at 54,000 ft above Hong Kong, still climbing at 700 ft/min with a fantastic view of China all the way to Canton.

On occasions, a pair of aircraft went on a weekend navigation trip to Clarke AFB in the Philippines. The USAF master sergeant, who was refuelling the Venom after our arrival, asked if the panel below the cockpit canopy was wood. He then went and knocked on the panel and called to his assistant on the fuel truck, "Hey, we've a wooden jet fighter here!"

On 24 May 1962, I had my last flight in a Venom, as the aircraft was officially retired from RAF service. The following month, the squadron re-equipped with the Hunter FGA.9. These were flown out from the UK to Hong Kong by 28 Squadron pilots, with a civilian engineering team waiting at Kai Tak to complete the acceptance by RAF engineers at the base. Whilst we were waiting for the Hunters to arrive, I was sent on to the jungle survival course at RAF Changi in Singapore, in June. After a week's training at Changi, we left to go upcountry into Malaya to spend eight days in the jungle. This was very exciting as we saw a few jungle animals such as monkeys, snakes, civet cats, a king cobra and all the insect life on the jungle floor. We slept in two-man bashas and single parachute hammocks during our treks through the jungle. The course finished with an escape and evasion exercise on Singapore island, where every house we passed had very noisy barking dogs. I managed to avoid being caught by the pursuing troops, thank goodness, and so did not have to go through an interrogation phase. I was back to Kai Tak and flying the Hunter on 28 June. The daily flying programme was very similar to the exercises carried out on the Venom, but with more formation flying, as the Hunters were much more serviceable than the Venoms.

Typhoon 'Wanda' arrived on 1 September 1962, and was one of the most intense tropical cyclones on record to hit Hong Kong. The whole station was battened down to withstand the storm and all aircraft that could not be hangered, were flown out to Taiwan or the Philippines. My station duty was in the squadron HQ, near the aircraft turnround pan and close to the sea front. It certainly was an extraordinary experience to see the extreme weather caused by the typhoon. Water was spurting through the door frames, complete windows were being blown out, and the rain reduced the visibility to twenty–thirty yards at times. Manhole covers were sucked out of the ground and blown away. When the eye of the storm passed over the airfield, and the wind and rain decreased to almost nothing, Dave Scott and I drove down to Kowloon to see the damage down there. It was mainly a few roofs off and lots of neon signs destroyed. We returned quickly as the other side of the typhoon came through in the afternoon. The main damage around the colony was caused by a storm surge more than seventeen feet above high tide in places. The following morning, on the weather check flight, we counted twelve large ships washed aground in the New Territories. Sha Tin airfield, in the New Territories just behind Lion Rock, was destroyed. It even had a large fishing junk washed up high and dry on the airfield.

1962 ended very sadly for 28 Squadron, as Dave Crichton was killed just after Christmas when his aircraft flew into Lion Rock. The cause of the accident was never established. In January 1963 I was detached for a week to fly with 20 Squadron at Tengah in Singapore. This was to give me experience in operating with a full squadron for the first time, as we only had a flight of five aircraft in Hong Kong,

Peter Headley leaves his aircraft in the hands of the ground crew. (AHB)

In May 1963, I returned to the UK to complete the PAI course at RAF Binbrook. I passed the course three months later and returned to 28 as a fully qualified PAI. When I returned in mid-August, I was very surprised to be immediately appointed close escort officer to a 20 Squadron pilot, who had been detached to Hong Kong to replace me whilst I was in UK. It appeared that whilst on a low-level navigation exercise, he had accidently flown into China, over the border near Lo Wu. He was waiting for the results of a Board of Inquiry, before returning to Singapore. (See Chapter 18.)

In September that year, the aircraft carrier HMS *Eagle* visited Hong Kong and 28 was invited to practise air-to-ground cannon and rocket attacks using the splash target towed by the carrier. This was also used by the carrier's Scimitar and Sea Vixen aircraft. The exercise area was thirty to fifty miles south of Hong Kong, in the South China Sea. This moving target was more fun and more demanding than our stationary target in Port Shelter. We also had a day out on the carrier, as there needed to be an RAF observer on the carrier bridge, when 28 Squadron aircraft were firing.

In October 1963, 28 had its annual squadron familiarisation flight to Shek Kong airfield, in the New Territories, to be hosted by the Royal Hong Kong Auxiliary Air Force for a dining-in night. This was held under marquees alongside the runway with a traditional Chinese menu for the dinner. It consisted of seven courses washed down by lots of brandy, I seem to remember. Some after-dinner games enlivened the proceedings. These included racing cars down the runway resulting in one car ending up driving through one of the marquees. Luckily, no one was hurt.

Later that month, Dave Scott and I flew a two-ship down to visit the USAF at Clark AFB in the Philippines. A good weekend allowed us to see a lot of the local area whilst being hosted by the F-102 squadron. No. 28 used to fly two or three of these trips each year, giving us some long-range navigation practice. This came in useful in 1964, when I had to deliver the command spare aircraft to Singapore, escorted by a Canberra from Tengah. The Canberra's navigation kit went u/s soon after leaving Labuan, so I had to lead it back to Tengah.

On 18 March 1964 I was detached once again to 20 Squadron at RAF Tengah, this time to operate from RAF Labuan, North Borneo, during the Confrontation. We carried out border patrols each day and occasionally escorted Beverley freighters carrying out parachute-supply drops. We also searched for lost SAS teams in the jungle near the Indonesian border and carried out FAC training. This detachment finished on 19 April and then I went to Udon Thani in north-east Thailand on another detachment with 20 Squadron. The airfield was close to the Laos border on the Mekong River. It was a Thai air force station with a squadron of North American T-6 Texans, and it also had an Air America detachment there, carrying out espionage and supply drops in Laos and Vietnam

No. 28 Squadron pilots with Peter Headley second from right. (AHB)

with C-46 aircraft. After a few enjoyable days exploring Bangkok and enjoying an air show by the Thai air force, I returned to Hong Kong.

Life in Hong Kong during the 1960s could not have been more different than the orderly, disciplined way of life of the RAF College at Cranwell that I had left just six months before arriving in the exciting and vibrant colony. As a young first tour pilot flying single-seat fighters in a totally new environment, and from a difficult airfield, it could not have been a greater contrast than the flying over the green fields of the UK or Europe with diversion airfields always available. Although the routine flying from Hong Kong was somewhat repetitive, the opportunity to operate with other units in the Far East was an exciting one and offered a steep learning curve.

My time in Hong Kong gave me unique experiences. The flying was demanding, different and exciting and presented problems that had to be professionally resolved. The opportunity to explore other fascinating places in the Far East was a bonus and the lifestyle and social life in the colony was something most people must spend a fortune to experience, and perhaps only on a two-week visit. It was an amazing introduction to life on an RAF squadron during a period when overseas service was part of a routine career pattern.

CHAPTER TWENTY-FOUR

HERCULES IN THE FAR EAST

STEPHEN WRIGLEY

I SUPPOSE THE STORY OF MY time in Singapore really starts when I was finishing my advanced flying training on the Varsity at RAF Oakington, Cambridgeshire. After a standard passage through RAF College Cranwell and RAF Oakington, I heard I was to be posted to fly the Andover. I wanted to go on to the C-130 Hercules, which were being purchased at the time by Harold Wilson's Labour government. A close friend and contemporary, John Mazurk, had been posted to the Hercules OCU but wanted to fly Andovers. We spoke to our instructor and by some magic, the nature of which I still do not understand even now, despite knowing more about the posting system than I did then, it was arranged for us to swap places. This decision had, I believe, a significance that I was only later to appreciate, as the Andovers were retired early and many of the pilots were sent to non-flying tours for a period. By the time I was a squadron leader and flight commander on a Hercules squadron, some of those Andover pilots were coming to the Hercules force as co-pilots.

Stephen Wrigley.

After advanced flying training, I was sent to No. 3 Hercules Course to train to be a co-pilot. All went reasonably well, and a posting to 48 Squadron in Singapore was to be my fate. However, having arranged to get married in October 1967, shortly before the end of the course, I was called to station HQ to be told to report to the air attaché in Bonn a few days before my planned wedding because I was to go to the German air force to fly the Noratlas. At a subsequent interview at Adastral House in London, having told the posters I would

A Hercules of 48 Squadron.

have to come back to the UK for a few days to get married, I was told that I could not possibly be sent to an exchange tour as an under 25-year-old since I would not qualify for marriage allowance. By this stage, my Singapore slot had already been allocated to a student on the course behind mine. But another stroke of good luck occurred as there were more changes and I got my slot in Singapore back. After this to-ing and fro-ing, ours is the only wedding invitation card that I know of that had an amendment slip stapled to it with a change of date. Such were the vagaries of my early service life after training.

After the Hercules course, we were all soon to have another lesson in the unpredictability of life in the transport force. No. 48 Squadron was established to have twelve Hercules C-130s and they were flown out to Singapore at the back end of 1967. While waiting, plans changed as the withdrawal from Aden was about to take place and a few of the new Hercules crews were held ready to evacuate personnel. A Britannia was prepared for us, and at least four Hercules crews were to proceed to Bahrain via Cyprus. The first attempt ended when the Britannia aborted take-off at RAF Lyneham. The number of qualified aircrew passengers who instantly realised what was happening was probably 100 per cent. The expectant looks of 'what next' could be seen in most faces.

The aircraft was soon repaired, and we made a routine arrival at RAF Akrotiri to be met by a warrant officer from the movements squadron. "Welcome to Cyprus gentlemen! You will be on the ground for a couple of hours and then be heading back to the UK." "Oh no!" we chanted: "Operational aircrew us. Aden withdrawal crews. We are going on to Bahrain." These exchanges went on for a few minutes and, of course, the WO was right. It was all a false start, and we departed for Lyneham. We eventually did arrive in Bahrain, and I see from my logbook that I flew Aden withdrawal sorties from 26 November to 30 November 1967.

The Hercules destined for 48 Squadron were flown out to RAF Changi in waves of four, and I arrived on 14 December 1967 to start what, in some ways, was probably my most enjoyable flying tour – lots of flying hours but not too much responsibility. It could accurately be summed up as we worked hard and played hard. But the vagaries of service life were to strike yet again right at the beginning of my tour. I had been told that my wife was not entitled to a passage to Singapore (again because I was under the magic age of 25) and she would have to request an indulgence flight. These were allocated when there were seats available not required for duty passengers or other entitled individuals. When she would get one was very much a matter of chance and would not happen for at least a month or two – especially in the run up to Christmas. On arrival in Singapore, I was so tired that despite the noise of the revelry in the officers' mess, I missed the arrival party. When I awoke the next morning, I eventually saw a note had been slipped under my door which told me that my wife was due to arrive in about forty-eight hours. I had no accommodation for her, little money, and a mini crisis on my hands. Fortunately, that service camaraderie that I so much appreciate stepped in. My captain at the time found us accommodation in the hotel where he was staying until he had a quarter. I don't know how he did it as it was full of similar personnel waiting for quarters, but perhaps the fact that he had lived in Singapore before helped in some way. Anyway, my new wife duly arrived just in time for Christmas, and several parties. As luck would have it, being a well-qualified nurse, she was able to quickly gain a role in the children's ward at the RAF Hospital in Changi as a locally engaged civilian. Although the pay was not much, it allowed us to live in a bungalow very close to Changi where many of the 'over 25s' were housed. Most of my 'under 25' colleagues had to live much further away from the base in accommodation that was much more modest.

The squadron's bailiwick was the Indian Ocean and the Pacific Ocean as part of the RAF's FEAF. There was a lot of flying with captains who had experience of the Hastings, Beverleys and other transport types and could teach us 'wet behind the ears' co-pilots a lot. The climate held its challenges for flying with monsoons, tropical storms, and thunderstorms, while also enabling us to water ski and enjoy swimming,

The elegant officers' mess at RAF Changi.

sunbathing and other sports. At RAF Changi, we also had the Far East Communications Squadron, 205 Squadron operating the Shackleton and a Meteor target-towing flight. And, of course, Changi was the transport destination for the Comets, Britannias, VC10s and Hercules that flew out from the UK. There was always a lot going on both operationally and socially. Fleet Air Arm squadrons would look in from time to time when a carrier was in the vicinity.

A lot of the flying from Changi was routine resupply to various bases and airfields. We undertook regular flights to Hong Kong and, initially we had an eighteen-hour crew duty day which allowed us to do a return flight to Hong Kong via Butterworth in Malaysia in one day. This eventually proved too much, and the crew duty day was reduced to sixteen hours. As a result, we started to position a slip crew at Hong Kong so that one crew flew up to Hong Kong, and the slip crew flew it back the same day. The first crew then had an overnight stay in Hong Kong, which was always welcome, and became the slip crew to fly back the day after arrival.

We also flew down to Australia and flights to Woomera, where all sorts of testing took place, are memorable. On one such trip, the navigator set up a descent to allow a visual arrival at Woomera but, at a great distance out, the captain said that he had the airfield in sight and the navigator could cease giving information. They landed and having taxied round to the parking area were met by an Australian in the proverbial bush hat who said something like: "if you guys want Woomera, it is fifteen miles south-east of here – this is Evetts Field." The story got back to Singapore before the crew arrived in Woomera.

We were a tactical transport squadron, so low-level flying, para-dropping and airborne resupply were all part of our remit. Most of the low-level flights went up into Malaysia where there were hills and mountains as well as jungle terrain, which made map reading interesting. Naturally, we did a lot of continuation training, which served young co-pilots well as we still had much to learn about operating, not just flying the aircraft. Captaincy potential was what the system was looking out for, and I am convinced that the quality of the captains on the squadron, and their combined experience, gave us role models in that regard.

Of course, like any squadron there were characters. One captain would come into the squadron operations room, where we tended to congregate first thing, and have a daily rant. It might be against the powers that be or some other underserving target of his angry tongue, but it was soon over and he felt better for it and we all had something to talk about for a while. On one trip with him, I remember saying that it was always good to have a beer after a long flight to relax. "You don't want a beer co-pilot," he growled, "what you need is a good book." It was only later that I gathered that he was something of a reformed drinker and, like a reformed smoker, they usually become the severest critics of their earlier habits.

His criticism of co-pilots could be withering, and probably lay in the fact that he always held it against us that he had to be a second pilot before becoming a co-pilot.

Flying in the Far East was always interesting and full of events and incidents, and not without danger. Several of my Cranwell entry (89) were in Singapore at the same time flying Lightnings, Hunters and Canberras. We lost two to accidents on Lightnings, a near wheels-up in a Hunter by a contemporary who went on to be our chief of the air staff, while a Canberra pilot had a problem with the control column which meant that the aircraft could not be controlled with it, and he flew on to Darwin and landed safely flying with only the trim controls. No mean feat, I believe. A month later he was interviewed by his AOC for damaging his aircraft on a mishandled take-off. Triumph and potential tragedy can be close neighbours in the flying world. Fortunately, he came out on the right side of lucky.

One accident that shows how fickle fortune can be happened to another Cranwell contemporary, who was flying Hunters at RAF Tengah. There was a big air exercise on the east coast of Malaysia in June 1968 involving air defence, ground attack and tactical resupply by 48 Squadron's Hercules. Many aircraft were airborne including several transport aircraft at the same time. The Hunter experienced an engine problem and the pilot ejected. As my Hercules was in the area returning to Changi having just dropped our load, we immediately converted to a search and rescue mission. As luck would have it, I was being 'comms checked' by the Wing AEO and he took over HF comms to HQ FEAF, while my captain and I concentrated on setting up a search pattern, taking charge of the scene of search and liaising with other aircraft while searching visually for a downed aircraft. Unfortunately, the Hunter pilot had got his north and south confused before ejecting, so much of the search was not in the right area. We did not locate him, although we investigated many small forest fires in the jungle while looking, and eventually we and the other aircraft involved returned to RAF Changi. We learnt later that he had landed on the beach and had called his base from a beach bar telephone, while sinking a few Tiger beers. Given the relatively shallow sea where the aircraft had crashed, it was allegedly too far out for land salvage and too close to the shore for sea salvage. Sadly, despite having been offered a slot in the Hercules world afterwards, he eventually retired from the air force probably because of the back injury he sustained during the ejection.

MAURITIUS MARCH–JUNE 1968

Work was not just about training and resupply flights. The Hercules Force has always been involved in military operations and, soon after our arrival in Singapore, we had an operational commitment in Mauritius. Early in 1968, several violent clashes occurred in various neighbourhoods of Port Louis shortly before the country declared independence on 12 March 1968. There was tension in the communities between Creoles and Indo-Mauritian Muslims, which apparently had its origins in gang warfare, and concerns over the nation's future following independence.

A company of the King's Shropshire Light Infantry was flown in from Singapore with helicopters and around 150 troops following a declaration of a state of emergency. I believe some UK police were also deployed. For crews on 48 Squadron, it meant several long flights of around ten hours, usually via the RAF staging post at Gan, in the middle of the Indian Ocean, to Mauritius with a direct Mauritius–Changi flight on return. Notwithstanding the troubles, we enjoyed the comfort of a beautiful hotel, a high-powered speedboat for water skiing, and the tales of the hotel manager who had been a colonial policeman with the role of judge, jury and perhaps hangman in various parts of colonial Africa. His wife took my captain and me to a Chinese gambling den one night where I was amazed at the amount of cash on display. I was also deeply anxious as I had the flight impress (a significant amount of cash) in a money belt around my waist. I was relieved to return to the hotel in the middle of the night with it still in my possession. (Transport crews carried significant amounts of cash to pay route allowances and other bills incurred during a task in a flight imprest. It all had to be accounted for properly and was always delegated to the co-pilot. Various stories

The RAF staging post at Gan in the Indian Ocean.

used to circulate about how such money was exchanged in the various bazaars of the Far East, but they remained, as far as I know, just stories.)

Back in the sixties of course, the internet and mobile telephones were not yet on the scene. Most of the young personnel did not even have telephones in the home. Contact with family and friends in the UK was usually by means of the lightweight airmail letter. On a Saturday, if free, we would often meet up at lunchtime downtown Singapore for a few beers. In one pub, there was a series of clocks showing times in various parts of the world and I often used to reflect that family in the UK were probably just getting up in the morning when we were into our third or fourth round of lunchtime drinks. I did make it home on one training flight but, owing to a delay en route, instead of a good twenty-four hours with my parents, who I had not seen in over a year, I arrived home after little sleep at around 10:00, went shopping with them for a lot of things we could not get in Singapore, got home, fell asleep for the whole afternoon to be woken at 17:00 by a fellow crew member ready to drive us back to RAF Lyneham. It was definitely a 'flying visit' from the other side of the world.

Sport, particularly water sports, trips into Malaysia, and parties took up much of our spare time. And there was always the unexpected. Driving to work one day parallel to the beach, I saw one of the squadron loadmasters thrashing around in the sea with some local Singaporeans. I vaguely wondered what he was doing at that hour but must have had something on my mind as I did not stop. I later learned that he had joined them in trying to capture a python. I am still rather glad I did not stop. There was also the Jungle Survival School and crews were required to do a three-day course in the jungle in Malaysia. Apart from the heat and humidity, and one tree falling on somebody who had tied a hammock to it, my main memory is that, on returning to our bungalow, my wife refused to let me in until I had taken my stinking (she said) clothes off outside before I was allowed in for a most welcome shower.

It is probably difficult to have a career in the RAF and not be touched by tragedy – and such events often happen early in one's service. I mentioned some aircraft accidents previously, but one sad car accident also occurred during my time on 48 Squadron. After meeting in the mess one Saturday evening, a group of us young co-pilots and navigators decided to go downtown Singapore for a satay meal on one of the many stalls there. While still on the station one of the cars overturned. All seemed OK apart from one, a young navigator, who was sadly killed. We were the next car on the scene, and my wife, a nurse, dealt with the situation. It hit us all very hard indeed. Looking back, I think that early experience helped us to cope with other situations during the rest of my time in the RAF. Service life can be fun, rewarding and at times exhilarating, but accidents and tragedy sometimes go hand in hand with it, as my subsequent career was to demonstrate.

ABBREVIATIONS

AC	Army Cooperation
ACMI	Air Combat Manoeuvring Instrumentation
ADC	Aide-de-camp
ADIZ	Air Defence Identification Zone
AEO	Air Electronics Officer
AGL	Above Ground Level
APC	Armament Practice Camp
ATAF	Allied Tactical Air Force
ATC	Air Traffic Control
AQM	Air Quartermaster
BUA	British United Airlines
BZ	British Zone
CCF	Combined Cadet Force
CENTO	Central Treaty Organisation
CFS	Central Flying School
CSE	Combined Services Entertainment
DZ	Drop Zone
FAC	Forward Air Control
FEAF	Far East Air Force
FGA	Fighter Ground Attack
FLOSY	Front for the Liberation of Occupied South Yemen
FOB	Forward Operating Base
FR	Fighter Reconnaissance
FTS	Flying Training School
HE	High Explosive
IGB	Inner German Border
IP	Initial Point
IRE	Instrument Rating Examiners
KD	Khaki Drill
LABS	Low-Altitude Bombing System
MEC	Middle East Command
MFPU	Mobile Field Photographic Unit
MOD	Ministry of Defence
MRT	Medium Range Transport
NBC	Nuclear, Biological, Chemical
NCO	Non-Commissioned Officer
NEAF	Near East Air Force
NLF	National Liberation Front
OCU	Operational Conversion Unit
ORB	Operations Record Book
P&SS	Provost and Security Services
PAI	Pilot Attack Instructor
PI	Photographic Interpreter
PI	Practice Intercept
QFI	Qualified Flying Instructor
QRA	Quick Reaction Alert
RIC	Reconnaissance Intelligence Centre
RSO	Range Safety Officer
SACEUR	Supreme Allied Commander Europe
SAM	Surface-to-Air Missile
SAR	Search and Rescue
SEATO	South-East Asia Treaty Organisation
TACEVAL	Tactical Evaluation
SATCO	Senior Air Traffic Control Officer
TFR	Terrain Following Radar
u/s	Unserviceable
USAF	United States Air Force
VHF	Very High Frequency
VSTOL	Vertical and Short Take-Off and Landing

INDEX